Environmental Policy

Environmental Policy clearly explains how the social sciences relate to environmental policy making, and how they can be used to achieve policies for a sustainable future. It deals with environmental policy making at institutional, national and international levels, and emphasises the solutions, as well as the problems.

Within the overall context of sustainable development, the book discusses the opportunities and constraints that environmental systems place upon the operation of human systems. The author suggests environmental policy as a potential way to modify the operation of human systems so that they function within environmental constraints. Key social scientific concepts (political, social and economic) are used to explain the background to the formulation and implementation of environmental policy.

Environmental problems, the role of human beings in creating them, sustainable development and how this concept relates to environmental policy are all introduced. The book then considers environmental policy formulation, implementation and evaluation within three specific contexts: the firm, the nation state and at an international level. It also reviews the place of economics, science and technology in environmental policy.

Environmental Policy is an accessible text with a multi-disciplinary perspective. Detailed case studies, drawn from a range of international examples, are used throughout to illustrate issues such as global warming, international trade, tourism and the human rights of indigenous peoples. It is well illustrated and includes chapter summaries and further reading.

Jane Roberts is Principal Lecturer in Environmental Policy at the University of Gloucestershire.

Routledge Introductions to Environment Series
Published Titles

Titles under Series Editors:
Rita Gardner and A. M. Mannion

Environmental Science texts

Atmospheric Processes and Systems
Natural Environmental Change
Biodiversity and Conservation
Ecosystems
Environmental Biology
Using Statistics to Understand the
  Environment
Coastal Systems
Environmental Physics
Environmental Chemistry

Titles under Series Editor:
David Pepper

Environment and Society texts

Environment and Philosophy
Environment and Social Theory
Energy, Society and Environment,
  2nd edition
Environment and Tourism
Gender and Environment
Environment and Business
Environment and Politics, 2nd edition
Environment and Law
Environment and Society
Environmental Policy

Routledge Introductions to Environment Series

Environmental Policy

Jane Roberts

 Routledge
Taylor & Francis Group

LONDON AND NEW YORK

First published 2004
by Routledge
11 New Fetter Lane, London EC4P 4EE

Simultaneously published in the USA and Canada
by Routledge
29 West 35th Street, New York, NY 10001

Routledge is an imprint of the Taylor & Francis Group

Typeset in Times by Keystroke, Jacaranda Lodge, Wolverhampton
Printed and bound in Great Britain by TJ International Ltd, Padstow,
Cornwall

British Library Cataloguing in Publication Data
A catalogue record for this book is available from the British Library

Library of Congress Cataloging in Publication Data
Roberts, Jane (Susan Jane)
    Environmental policy / Jane Roberts.
        p. cm. – (Routledge introductions to environment series)
    Includes bibliographical references and index.
    1. Environmental policy. I. Title. II. Series.

GE170.R59 2004
363.7′0561—dc21   2003008789

ISBN 0–415–19885–2 (hbk)
ISBN 0–415–19886–0 (pbk)

Contents

List of figures — *vi*

List of tables — *vii*

List of boxes — *viii*

Series editor's preface — *ix*

Acknowledgements — *xii*

Introduction — 1

Chapter 1   So what's the problem? — 6

Chapter 2   The roots of environmental problems — 40

Chapter 3   Sustainable development and the goals of environmental policy — 66

Chapter 4   Science and technology: policies and paradoxes — 90

Chapter 5   Corporate environmental policy making — 118

Chapter 6   Environmental policy making in government — 139

Chapter 7   International environmental policy making — 167

Chapter 8   Environmental economics — 189

Chapter 9   Making policy for the environment – and for people — 217

References — 219

Index — 230

Figures

1.1	The fish resource cycle	9
1.2	The copper resource cycle	10
1.3	The hydrological cycle	12
1.4	Bio-accumulation of pesticide residues	23
1.5	World population growth, actual and projected, 1950–2050	28
2.1	Maslow's hierarchy of needs	43
3.1	Reference case scenario from *The Limits to Growth*	70
3.2	Scenario from *The Limits to Growth* assuming unlimited resources	71
3.3	Weak and strong sustainability	82
4.1	Risk and precaution	100
4.2	Preventative and end-of-pipe approaches in a manufacturing process	111
4.3	The waste management hierarchy as a decision-making framework	114
5.1	The stages of an environmental management system	129
6.1	The policy-making process at government level	140
6.2	Downs's issue attention cycle	141
6.3	Interest groups and representation	143
6.4	Regulatory and economic systems of pollution control	158
7.1	Ecological footprint, by region, 1996	175
7.2	Debt in the context of GNP and exports, 1980 and 1995	179
8.1	Supply and demand curves	190
8.2	The effect of changes in supply characteristics on equilibrium prices and quantities	192
8.3	The effect of changes in demand characteristics on equilibrium prices and quantities	192
8.4	Price inelasticity of demand	193
8.5	The effect of a pollution tax on price and quantity	199
8.6	The effect of a hypothecated pollution tax on price and quantity	204

Tables

1.1 Renewable and non-renewable resources in Great Britain   11

1.2 Global reserve/production ratios for some non-renewable resources   13

1.3 Examples of environmental sinks   16

1.4 Trends in total fertility rate and population growth by country and region   29

1.5 Arguments for conserving species   32

2.1 Attributes of successful common property regimes   55

3.1 Mid-1990s consumption figures and environmental space targets for some key resources   85

4.1 Attributes and scale descriptors for framing policy problems in sustainability   91

5.1 The SIGMA management framework   134

5.2 The Natural Step's four system conditions   135

6.1 Types of policy networks: characteristics of policy communities and issue networks   150

7.1 International bodies and agreements relevant to international development, trade and environment negotiations   177

8.1 The future value of £100 at different discount rates   194

8.2 The present value of £100 in future years for a range of discount rates   195

8.3 The cost of environmental degradation in Nigeria   210

8.4 Indicators of the quality and sustainability of urban environments   212

Boxes

1.1	Pesticide waste systems	19
1.2	The carbon cycle and global warming	21
1.3	Sea-level rise and coastal erosion	25
2.1	Easter Island: statues and status	45
2.2	The Framework Convention on Climate Change	51
2.3	Coronation Hill	56
3.1	Zimbabwe's CAMPFIRE project	76
3.2	Forest management in Finland	78
3.3	Forest management in Madagascar	80
4.1	Bovine spongiform encephalopathy	95
4.2	The Green revolution	103
4.3	The Severn tidal barrage	107
4.4	Waste management in Hamburg	112
5.1	Asbestos: Turner & Newall	121
5.2	Du Pont and the CFC phase-out	122
6.1	Friends of the Earth and Greenpeace	146
6.2	The non-fossil fuel obligation	154
6.3	Policy instruments for population control	162
7.1	Controlling sulphur emissions in Europe	169
7.2	Senegal: fish, trade and sustainability	171
8.1	Balancing inter-generational costs: US nuclear power	196
8.2	The UK limestone and landfill taxes	200
8.3	Local exchange trading systems	214

Series editor's preface
Environment and Society titles

The modern environmentalist movement grew hugely in the last third of the twentieth century. It reflected popular and academic concerns about the local and global degradation of the physical environment which was increasingly being documented by scientists (and which is the subject of the companion series to this, Environmental Science). However it soon became clear that reversing such degradation was not merely a technical and managerial matter: merely knowing about environmental problems did not of itself guarantee that governments, businesses or individuals would do anything about them. It is now acknowledged that a critical understanding of socio-economic, political and cultural processes and structures is central in understanding environmental problems and establishing environmentally sustainable development. Hence the maturing of environmentalism has been marked by prolific scholarship in the social sciences and humanities, exploring the complexity of society-environment relationships.

Such scholarship has been reflected in a proliferation of associated courses at undergraduate level. Many are taught within the 'modular' or equivalent organisational frameworks which have been widely adopted in higher education. These frameworks offer the advantages of flexible undergraduate programmes, but they also mean that knowledge may become segmented, and student learning pathways may arrange knowledge segments in a variety of sequences – often reflecting the individual requirements and backgrounds of each student rather than more traditional discipline-bound ways of arranging learning.

The volumes in this Environment and Society series of textbooks mirror this higher educational context, increasingly encountered in the early twenty-first century. They provide short, topic-centred texts on social science and humanities subjects relevant to contemporary society-environment relations. Their content and approach reflect the fact that each will be read by students from various disciplinary backgrounds, taking in not only social sciences and humanities but others such as physical and natural sciences. Such a readership is not always familiar with the disciplinary background to a topic, neither are readers necessarily going on to further develop their interest in the topic. Additionally, they cannot all automatically be thought of as having reached a similar stage in their studies – they may be first-, second- or third-year students.

The authors and editors of this series are mainly established teachers in higher education. Finding that more traditional integrated environmental studies and specialised texts do not always meet their own students' requirements, they have often had to write course materials more appropriate to the needs of the flexible undergraduate programme. Many of the volumes in this series represent in modified form the fruits of such labours, which all students can now share.

Much of the integrity and distinctiveness of the Environment and Society titles derives from their characteristic approach. To achieve the right mix of flexibility, breadth and depth, each volume is designed to create maximum accessibility to readers from a variety of backgrounds and attainment. Each leads into its topic by giving some necessary basic grounding, and leaves it usually by pointing towards areas for further potential development and study. There is introduction to the real-world context of the text's main topic, and to the basic concepts and questions in social sciences/humanities which are most relevant. At the core of the text is some exploration of the main issues. Although limitations are imposed here by the need to retain a book length and format affordable to students, some care is taken to indicate how the themes and issues presented may become more complicated, and to refer to the cognate issues and concepts that would need to be explored to gain deeper understanding. Annotated reading lists, case studies, overview diagrams, summary charts and self-check questions and exercises are among the pedagogic devices which we try to encourage our authors to use, to maximise the 'student friendliness' of these books.

Hence we hope that these concise volumes provide sufficient depth to maintain the interest of students with relevant backgrounds. At the same time, we try to ensure that they sketch out basic concepts and map their territory in a stimulating and approachable way for students to whom the whole area is new. Hopefully, the list of Environment and Society titles will provide modular and other students with an unparalleled range of perspectives on society-environment problems: one which should also be useful to students at both postgraduate and pre-higher education levels.

David Pepper

May 2000

Series International Advisory Board

Australasia: Dr P. Curson and Dr P. Mitchell, Macquarie University

North America: Professor L. Lewis, Clark University; Professor L. Rubinoff, Trent University

Europe: Professor P. Glasbergen, University of Utrecht; Professor van Dam-Mieras, Open University, The Netherlands

Acknowledgements

I have built up many debts of gratitude during the (too) long gestation of this text. First, I must thank my editors at Routledge, Sarah Lloyd and Andrew Mould, for their help and patience. Colleagues at the University of Gloucestershire assisted in many ways. Without the moral support of Gerry Metcalf, Barbara Hammond, Carolyn Roberts and Stephen Owen completion would even now be awaited. Margaret Harrison, John Powell and Martin Spray were kind enough to review some chapters, but are exonerated from responsibility for any errors that remain. The comments on a draft manuscript of three anonymous reviewers and Professor Stephen M. Meyer of Massachusetts Institute of Technology were also helpful. Kathryn Sharp and Trudi James are thanked for their patient and careful preparation of the figures. At home, Chris, Hazel and Anna put up with a lot, so thank you.

Thanks also are due to the students and graduates of the Environmental Policy degree at the University of Gloucestershire (and its former incarnation, Cheltenham and Gloucester College of Higher Education) for teaching me how to teach (an on-going process!); keeping me sharp by asking the right questions; and keeping in touch, so that I know the large and small differences they are making to the world in their working lives. Rachel Bridgeman is thanked particularly for her permission to use the photograph in Box 2.1.

Permission to reproduce the following figures is also gratefully acknowledged: the Club of Rome for Figures 3.1–2, the Finnish Forest Industries Federation for the figure in Box 3.2, the Intergovernmental Panel on Climate Change for the figure in Box 1.3 and the United Nations Population Fund for Figures 1.5 and 7.1. Figure 7.2 and those in Box 4.2 are Crown copyright material, reproduced with the permission of the Comptroller of HM Stationery Office and the Queen's Printer for Scotland.

J.R.

this goodly frame, the earth, seems to me a sterile promontory; this most excellent canopy the air, look you, this brave o'er-hanging firmament, this majestical roof fretted with golden fire – why, it appeareth no other thing to me than a foul and pestilent congregation of vapours.

(Hamlet, Act 2, Scene 2)

Introduction

In one often quoted metaphor (Cunningham 1963) policy is likened to an elephant, bringing to mind the Indian folk tale. Several blind men are led to an elephant and invited to describe what it is that they are feeling. As each is touching a different part of the animal (flank, tail, trunk, leg, tusk etc.) they argue about the nature of the beast. This analogy seems amusing and true to the experienced policy analyst yet it is singularly unhelpful to the beginner. A much better place to start is with the dictionary definition of policy:

> **policy** (n.) Political sagacity; statecraft; prudent conduct, sagacity; craftiness; course of action adopted by government, party, etc.
>
> (*Concise Oxford Dictionary*)

This definition suggests that 'policy' is a wise course of action and that the word is often used to describe the principles underlying actions undertaken in the political arena. Thus, at its most basic level, 'policy making' means developing the principles which will determine such a course of action.

Textbook definitions of policy are often focused at the governmental level. This is helpful when considering the policy processes of central government, where policy is as much a process as a product. But the concern of this book is simultaneously narrower and broader than that of most politics textbooks. The focus here is on environmental policy, a specialist area of concern. And central government policy making is only a part of the story – of equal interest are the levels of international and organisational policy making.

The working definition of policy used in this book is that *policy is a set of principles and intentions used to guide decision making*. This has the advantage that it is easily understandable and can be meaningfully applied to each level of decision making, from the individual to the United Nations. Thus, it may be my policy to reduce the environmental impact of my fuel consumption provided I can do so without undue cost or inconvenience. These principles guide decisions I take on domestic energy use and personal transport. I will switch off lights when I leave a room; I will cycle to work rather than drive if it is not raining. Similarly, it could be the policy of a certain government to decrease the environmental impact of waste disposal, where it is cost-effective to do so. Actions and targets resulting from this policy might aim to minimise the amount of waste produced and/or to increase the recycling of certain materials in the domestic waste stream by a given amount over a specific time period.

This very simple definition can be adapted into the definition of environmental policy used in this book: *environmental policy is a set of principles and intentions used to guide decision making about human management of environmental capital and environmental services.*

Why is environmental policy important?

Policy making is a web-like and multi-layered phenomenon which occurs at every level of human organisation from the individual to confederations of nation states. Increasingly, policy makers are being forced to focus upon the effects that human activities are having on the physical and biological systems of planet Earth. The United Nations Conference on Environment and Development, also known as the Earth Summit, which was held at Rio de Janeiro in 1992, attracted top-level representatives from 178 countries and was the largest international conference ever held, demonstrating the significance that environmental problems were assuming by the end of the twentieth century. An action plan aiming to reconcile economic development and environmental protection, Agenda 21, was agreed at the Earth Summit and was reviewed when the conference reconvened, ten years on, in Johannesburg in August 2002.

Agenda 21 calls for actions, not only by national governments, but also by local authorities, firms, voluntary organisations, communities and individuals. Environmental problem solving is a skill which is now required by those responsible for policy at every level of organisation, but environmental decision making cannot take place outside the wider context of economic and social responsibility inherent in the concept of sustainable development.

The upsurge in attention given to the environment is a result of mounting scientific evidence showing that present and projected patterns of economic activity are causing such severe environmental damage as to threaten their continuation. Scientific understanding of the natural environment has advanced greatly in recent decades. However, the complex manner in which physical and biological systems operate and interact means that, for many environmental problems, detailed scientific understanding of the relationship between causes and effects is an extremely challenging goal, especially given the long time scale (decades, centuries or even longer) over which some human effects on the environment are manifested.

However, if environmental problems are to be successfully resolved, rather than merely understood, knowledge of the scientific laws which govern the behaviour of natural systems needs to be complemented with insight into the social sciences. For example, the relationship between belief systems and environmental attitudes held by different societies is an important factor in determining the definition and resolution of environmental problems. The disciplines of politics and economics describe the principles which have been shown to underlie decision making in communities, organisations and nations. If human activities are the cause of environmental problems (and because it is often the threat of disruption to these activities which motivates the search for a resolution of these problems) then it is essential for environmental policy makers to understand the workings of human systems at least as well as they understand how environmental systems operate.

If environmental policy makers are to be successful environmental problem solvers, therefore, they need to bring a multi-disciplinary perspective to bear. They must be able to understand the significance of what scientists can tell them, yet also be able to use a range of social science methods to explain and analyse the causes of environmental problems, and the barriers to their solution, which lie within human societies. This book introduces, chapter by chapter, the necessary diverse range of approaches, making the links between these clear as they arise.

The structure of the book

Chapter 1 reviews the demands that humans make on the environment and how these can generate environmental problems. It establishes that the purpose of environmental policy is to change human behaviour – to make people act in ways which do not generate environmental problems, or which generate problems of lesser significance than was previously the case. Effective environmental policies are essential if progress towards sustainable development is to be made.

In Chapter 2 causes of such behaviour are analysed, for example Hardin's tragedy of the commons model, which suggests that over-exploitation of environmental capital is inevitable – unless 'mutual coercion, mutually agreed upon' can be adopted. Mutual coercion would necessarily take the form of policy. Desired outcomes would be developed and agreed upon by the group of commoners and then changes in activities agreed in order to achieve these outcomes. Chapter 3 introduces the concept of sustainable development as a potential goal of environmental policy, as well as more limited goals such as Best Practicable Environmental Option. None of these goals will be reached by chance or without intervention to change the ways in which people use environmental capital and services.

Chapter 4 examines the ways in which science and technology assist and impede policy makers in pursuit of these goals. As well as making use of scientific knowledge about environmental problems, therefore, policy makers need also to understand the limitations of scientific evidence and the nature of scientific uncertainty, especially when predictions of negative events in the far distant future might seem to justify a costly course of action in the here and now.

Chapter 5 explores the changing role of environmental policy within the organisation. In this and the next two chapters policy is introduced as a way of changing the behaviour of people, organisations and governments. It will be seen that policy as a concept cannot be disentangled from the context within which it is formulated and implemented. Because this context will have different features at governmental and at organisational levels these are considered separately, national in Chapter 6 and international in Chapter 7.

Throughout preceding chapters, the science of economics emerges as significant, either as a barrier to effective policy making, or as a potential tool for policy analysis or even implementation. Chapter 8 explores some of the barriers and the extent to which they can be overcome through the disciplines of environmental economics and ecological economics.

Chapter 9 pulls together the implications for policy makers of previous chapters, suggesting that the environmental policy 'toolkit' will be invaluable for decision makers in the twenty-first century seeking to reconcile the twin necessities of development and environmental protection.

The case studies

No text could cover more than a proportion of scholarship in this subject area and the treatment here is necessarily selective. Although the emphasis on multi-disciplinary problem solving means that this is not an 'issues' book, some important environmental issues are referred to in the text, using boxes. These short case studies are referenced to the academic literature so that students can choose to examine these in more detail where appropriate. They have been deliberately chosen to illustrate themes from the chapters and discussion points at the end of each box allow further drawing out of key concepts.

This way of exploring environmental policy is very different from the issue-by-issue structure that textbooks written for environmental studies courses often adopt. However, it is hoped that by adopting this approach *Environmental Policy* is able to offer a broad cross-disciplinary perspective on how some of the most important policy questions of the twenty-first century might be resolved.

❶ So what's the problem?

- The concepts of environmental capital and environmental services
- The issues of resources, waste and pollution, population growth, biodiversity and quality of life illustrating these concepts
- A definition of 'environmental problem' and its application
- The relative roles of natural and human factors in causing environmental problems
- The potential of environmental policy to prevent, diminish or solve environmental problems

The major environmental issues

Look around you.

Everything that you can see is either part of the environment or has been produced from resources that were extracted from the environment. Without an environment capable of providing the physiological resources of air, water and food human beings could not even have evolved. The global economy, which at the beginning of the twenty-first century was sustaining more than 6 billion human beings, is utterly dependent upon a stream of raw materials. Whether these are animal, vegetable or mineral in nature, their origin is environmental.

However, resource provision is only one aspect of the environment on which humankind is dependent. Air, water and land act as the necessary sinks for the wastes that are the inevitable products of the processes that demand resources. People use the environment to procure shelter, safety, aesthetic pleasure and spiritual sustenance. Each of these uses can be thought of as an *environmental service*: a service that the environment provides for the individuals who comprise the human race.

Conceptualising the environment in terms of its ability to service the human race is an approach increasingly used by environmental policy makers. Attributes of the environment can be thought of as environmental capital capable of providing services which people can use. Thus a river is environmental capital to the extent that it provides environmental services, for example, water for abstraction and fish to be eaten. Other environmental services the river might supply are to receive and carry away storm water and sewage from human habitations, and to represent a recreational and leisure resource as pleasant place for people to enjoy.

The environmental capital and services concepts are borrowed from economics, where financial capital is money which has been invested to produce a stream of income from interest or dividends. The concepts are useful in the definition and characterisation of environmental problems and are therefore used to underpin the analysis in this (and later) chapters of some of the key environmental issues which are perplexing policy makers at the beginning of the twenty-first century: resources; pollution and waste; population growth; biodiversity; and quality of life.

Resources

The term 'resource' is used to describe:

- material resources of use to individuals and society;
- flows of energy which can be harnessed for useful purposes;
- attributes of the environment that contribute something of value.

The allied concepts of usefulness and value are therefore key in the definition of resources and are, of course, culturally determined. Even the ways in which the basic resource needs for food, water and materials to construct shelter and warm clothing are met vary between cultural groups.

Examples of material resources are minerals, such as metal ores or stone for buildings; or agricultural or forestry products. Usually, when the term 'resources' is used, it will be a reference to material resources such as these which have clear economic value and can be accounted for in terms of weight or volume. When coal or uranium is extracted from the ground, this is an example of a mineral resource being mined in order to provide energy. The primary resource we are concerned with is a material substance.

It is possible to analyse the use of material resources and production of wastes in human economies by looking at the complete life cycle of a resource, from its environmental cradle to its environmental grave. Such a system is called a resource cycle.

Consider the resource cycles depicted in Figures 1.1–2. Note that primary resources are those extracted directly from the environment, whilst secondary resources are obtained from materials which have already entered the resource cycle, e.g. by recycling. It can be seen that the components of these systems are:

- extraction of the primary resource from the environment;
- concentration, refining and purification of the resource;
- use of the resource to manufacture economically useful goods;
- use of the goods within the human economy;
- designation of the goods, or their by-products, as wastes at the end of their usefulness;
- possible recovery of secondary resources, i.e. materials or energy, from the waste materials;
- disposal of the waste materials;
- assimilation of the waste materials into environmental sinks.

Note that the resource cycle diagrams give no information about the relative locations of the component processes, nor the lengths of time that elapse between the processes. Also missing from the diagram are the other resources which are needed to extract, use and dispose of the resource and resultant waste materials – for example, the resources needed to produce energy to power these processes.

Energy flows can also be regarded as resources. When technological devices or buildings are designed to capture energy from the environment – for example, wind turbines or solar heated buildings – the primary resources are forms of energy, not materials.

The word 'resource' is also used to describe attributes of the environment. The term 'land resource' is used to describe the hectarage of land available for a particular purpose, for example arable crops, grassland for grazing or moorland for recreation. Rivers and oceans are also resources, providing fish and other foodstuffs. Whereas some resources of this type are of direct economic use because they are the source, for example, of inputs to agriculture or manufacturing, other environmental attributes have value of a different kind. For example, local communities often conceive open space in a city as a resource, yet it creates no tangible economic outputs. The contribution that this type of environmental service makes to the quality of life is discussed later in this chapter.

Flow and stock resources

Resources are often classified into renewable (or flow) resources and non-renewable (or stock) resources. For renewable resources the rate at which natural cycles produce the resource is of the same order, or faster than, the rate at which the resource is consumed, thus maintaining

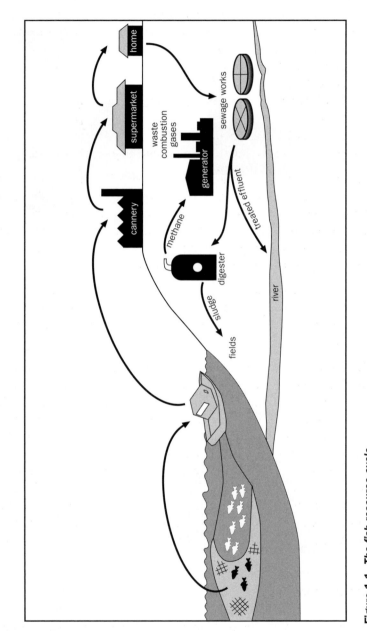

Figure 1.1 The fish resource cycle

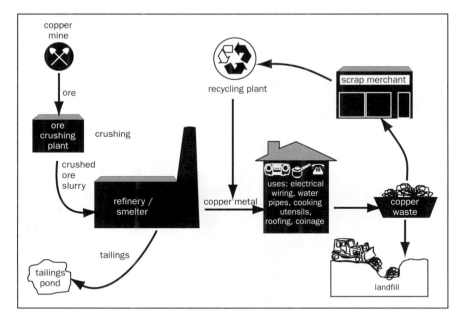

Figure 1.2 *The copper resource cycle*

environmental capital. For non-renewable resources the rate of production of the resource is much slower than the rate at which the resource is consumed, so that environmental capital is inevitably depleted.

Table 1.1 gives examples of renewable and non-renewable resources. For some resources the distinction is clear. Fossil fuels (such as oil, coal and natural gas) were formed by biological and geological processes that have taken place since the Carboniferous period, 300 million years ago. Given this extremely long cycle of generation, fossil fuels are non-renewable resources. Similarly, some resources are clearly renewable. Energy from the Sun falls upon the Earth at a rate equivalent to the heat production of 173 million million one-bar electric fires. It is clearly impossible to 'use up' this resource, as human activity can have only a minor effect on its rate of arrival.

For some other resources, however, the boundary between renewable and non-renewable is less distinct, as Table 1.1 shows. Paper is made from wood pulp, which is a renewable resource – but only if forestry is managed so that timber is regenerated at the same rate that it is harvested (see Box 3.2). In this case, renewability becomes dependent on environmental management – maintaining a balance between the rate of use and the rate of regeneration. If this does not occur then the resource

Table 1.1 *Renewable and non-renewable resources in Great Britain*

Resource	Typical time span since resource was formed (years)	Renewable?
Limestone	320 million	No: geological regeneration processes are many times longer than a human generation
Coal	300 million	No: geological regeneration processes are many times longer than a human generation
Lignite	35 million	No: geological regeneration processes are many times longer than a human generation
Peat	100,000	No: geological regeneration processes are many times longer than a human generation
Oak timber	100	Marginal
Spruce timber	40	Yes, provided replanting allows regeneration
Meat	1	Yes, provided husbandry allows breeding and regeneration
Fruit and vegetables	<1	Yes, potentially
Freshwater	0 – up to 1 million	Yes

becomes depleted, as is the case with some fish populations which have become depleted through over-exploitation, for example those of the north-west Atlantic (MacGarvin 2002; see also Box 7.2).

Sometimes it is not just the quantity, but also the quality of a resource that is important. Water is a good example here. Water as a chemical (H_2O) is abundant on the Earth – 71 per cent of the globe's surface is covered by oceans. Fresh water (or non-saline water), suitable for drinking, agricultural and industrial uses is, however, a very small proportion of this and most of it is locked up in ice caps in the polar areas. The hydrological cycle of evaporation from the oceans and precipitation over the land not only resupplies water to land areas, but purifies the water in the same way as chemical distillation. So when fresh water is discussed as a 'renewable resource' it is the quality of the water which is said to be renewed, not the quantity.

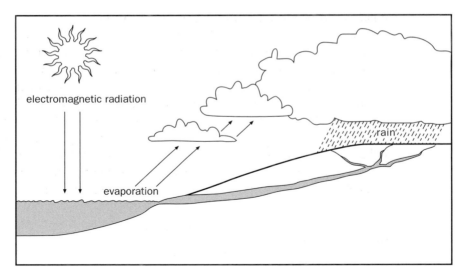

Figure 1.3 *The hydrological cycle*

Resource depletion

An appropriate balance between use and regeneration is therefore an important goal for resource management. If resources are used at rates greater than the rate of regeneration, the time will come when the supply of the resource will become scarce. How soon this happens will depend on the size of the initial stock of the resource and on the relative rates of use and regeneration. Renewability is no guarantee that a resource will be maintained and non-renewability does not necessarily imply future scarcity.

An isolated population of fish in a lake may be extinguished by overfishing in one or two seasons if too great a percentage of the adult fish is harvested, although restocking of the lake would be a possibility in these circumstances. In contrast, some non-renewable resources are used up so slowly that they are unlikely ever to be extinguished. Silica, a raw material used in glass making and some electrical components, is an example. Silica is extremely abundant – silicate minerals make up the bulk of the Earth's mass. Globally, glass making could continue virtually indefinitely before the silica resource was exhausted.

For any resource, the reserves are those parts of it which have been demonstrated to exist and which it would be possible to extract or use at a current economic cost. For mineral resources, global reserves are likely to be a very small proportion of the total resource that actually exists and new reserves are constantly being identified.

When stocks of non-renewable resources are calculated, this is done using reserve/production ratios, calculated by dividing the quantity of the resource in reserves by the current annual production. This will give the number of years that the reserves would last, assuming no new prospecting and no change in the rate of use. It is important that the calculation is undertaken at an appropriate spatial scale, be it local, regional or global. In the UK, for example, this ratio is established at regional level for sand and gravel reserves, which are low value, bulky and therefore relatively expensive to transport. For a valuable material such as gold, with reserves that are unevenly spread between countries, a global scale would be appropriate to perform the calculation.

$$\text{Reserve/production ratio} = \frac{\text{Quantity of reserves}}{\text{Current annual production}}$$

Table 1.2 shows the global reserve/production ratios for some mineral resources that are important to human lifestyles. It is clear that, although some reserves are in abundant supply, others are seemingly close to exhaustion. In the 1960s and 1970s the future availability of such resources was seen as one of the most important environmental issues (see Meadows *et al.* 1972 and also Chapter 2). Concern is now much less than it was as the then predicted resource shortages have mostly failed to materialise. This is despite the fact that the rate of utilisation of many non-renewable resources has increased dramatically since the early 1970s.

Table 1.2 *Global reserve/production ratios for some non-renewable resources*

Resource	Year	Reserves	Annual production	Reserve/ production ratio
Coal[a]	2001	984×10^9 tonnes	4.56×10^9 tonnes	216
Gas[a]	2001	155×10^{12} m³	2.46×10^{12} m³	63
Oil[a]	2001	143×10^9 tonnes	3.58×10^9 tonnes	40
Copper[b]	2000	340×10^6 tonnes	12.9×10^6 tonnes	26
Potash[b]	2000	8.4×10^9 tonnes	25×10^6 tonnes	336
Sulphur[b]	2000	1.4×10^9 tons	57×10^6 tons	24.6

Sources: a BP (2002); b Wagner *et al.* (2002)

Note: One (metric) tonne is the equivalent of 1.1023 (US) tons.

The reasons for this are partly economic and partly technical. If a resource becomes scarce, its price tends to rise in line with the law of supply and demand. Firms operating in the market place predict the coming shortage and company geologists are dispatched to prospect for new reserves, which, once verified as technically and economically viable, can be added to the reserves total. This process has been aided by new technological developments in mineral prospecting, such as remotely sensed satellite imaging and deep-sea drilling technology, which have extended the boundaries of search areas into previously inhospitable or unmapped terrain and allowed the identification of previously inaccessible reserves.

Once located, these new reserves may be as cheap to extract as the existing reserves, in which case the resource will remain in a supply about as plentiful as before on the world market, and at about the same price. Alternatively, the new reserves may be more expensive to extract. In the case of mineral ores, for example, the new deposits may be of poorer quality. The sites at which the newly found reserves are located may be inaccessible: geographically remote from the existing transport infrastructure, or technically difficult to extract (e.g. under the sea bed). This rise in production costs may be offset by improvements in the technology of extraction, but, all other things being equal, a resource that is getting scarcer will also be getting more expensive. This has two possible effects. First, the higher value of the resource will act to encourage further prospecting, and then more deposits will qualify as economically viable reserves at the higher price. Thus scarcity can be self-rectifying, within limits. As resources become more expensive to extract, however, it is likely that the energy inputs and waste outputs from the extraction process will be increasing, causing greater environmental impact for each unit of production.

Second, demand for the resource may diminish as the price rises and consumers are less willing to pay. One reason for this may be resource substitution, when alternative and cheaper materials are found to replace the original: plastic piping for domestic plumbing instead of copper, for example. Higher resource prices can also encourage recycling of materials, leading to smaller demand for the primary resource.

Paradoxically, it is more often the case that renewable, rather than non-renewable, resources are in danger of becoming scarce. Fisheries and forests in many parts of the world are being depleted at a far greater rate than they can replenish themselves, whilst the reserve/production ratios of fossil fuels and other minerals, on a global scale, remain steady or are even increasing. Scarcity of non-renewable resources, where it exists, is at the local level and often associated with poverty and inequitable distribution. This does not mean that concern about the continuing increasing use of non-renewable resources is misplaced, however, for three reasons.

First, the recent rapid developments in the technologies of mineral prospecting and extraction must, eventually, tail off. Once remote sensing can enable the detection of mineral resources anywhere on the earth's surface there will be nowhere else to look (except, perhaps, outer space). Improvements in the technology of mining and refining ores will show similar diminishing returns (McLaren *et al.* 1998: 209–11).

Second, mineral extraction can be environmentally damaging and energy-intensive. It is often the case that recycling waste products to produce secondary resources will save environmental damage and energy as well as displacing primary resources. Life cycle analysis and the waste hierarchy (introduced in Chapter 4) can be used to assess the relative merits of utilising primary and secondary resources.

Third, it can be seen that a further potential advantage of recycling is a reduction in the amount of waste exported back to the environment for disposal. However many times an object is reused, or the materials in it are recycled, the eventual fate of all the material resources extracted from the environment and used in the human economy is that they are returned to the environment as wastes, as Figure 1.1 shows. Over the last few decades, the major focus of concern for environmental policy makers has switched from resource depletion to waste production and its environmental impacts.

Waste and pollution

The ability of the environment to receive wastes is the second important set of environmental services considered in this chapter. The World Health Organisation definition of waste is 'something which the owner no longer wants at a given place or time and which has no current or perceived market value' (WHO 1971). When a part of the environment is used for the disposal of such valueless material it is called a sink, and is, of course, a form of environmental capital providing the service of waste disposal. Examples of environmental sinks are given in Table 1.3 and Boxes 1.1–2.

Waste can be classified along the same lines that were used for resources in Chapter 1. Note that the subjective determination of value is as important for the definition of wastes as it is for resources and these

Table 1.3 Examples of environmental sinks

Resource	Properties of the resource	Waste(s)	Properties of waste	Environmental sink
Coal	Coal is a solid fuel, formed by geological processes over hundreds of millions of years from prehistoric vegetation	Carbon dioxide	Greenhouse gas	Atmosphere
		Sulphur dioxide	Gas: reacts with atmosphere moisture to form acid rain, mist and snow	Atmosphere; fresh or marine waters; soils
		Nitrogen oxide	Gas	Atmosphere
		Ash	Solid	Landfill. Although ash from large coal furnaces may be used to make building blocks these will ultimately be disposed of to land when the building they were incorporated into has been demolished
Wood pulp (paper)	Paper is an organic material, rich in carbon. In a landfill it may decompose into a mixture of carbon dioxide and methane, or it may remain largely unaltered chemically. If incinerated, the main waste products are carbon dioxide and ash	Paper	Solid	Landfill

Material	Description	Product		
		Carbon dioxide	Greenhouse gas	Atmosphere
		Methane	Greenhouse gas	Atmosphere: eventually oxidised to carbon dioxide
		Ash	Solid	Landfill: some soluble salts may be leached into water courses
Copper	Solid metal: may react chemically to form compounds	Copper metal and copper compounds	Mostly solid, although some copper compounds are water soluble. Potentially toxic	Landfill: some soluble compounds may be leached into water courses
Uranium	Mildly radioactive element used as a fuel when electricity is generated using nuclear power	Radioactive waste: can be classified into high-level, intermediate-level and low-level waste (see Box 8.1). During irradiation in a nuclear reactor, uranium undergoes a series of fission reactions to produce a mixture of new elements	The combined radioactivity of this mixture is much greater than the radioactivity of the original uranium	Some radioactive wastes are discharged to the atmosphere (gases) or to water (soluble compounds). Most, however, are either disposed to specially designed and managed landfill sites, or are held in storage awaiting their eventual disposal to underground repositories

classifications are phrased to demonstrate this subjectivity. Wastes are:

- material resources that have been extracted from the environment and are now deemed to be of no further use to individuals and society;
- flows of energy resulting from human activities which it is decided are not worth using further (e.g. 'waste heat' from power stations);
- attributes of the environment that are not valued.

The term 'waste' can be used, not only to describe solid objects, such are as found in domestic dustbins and industrial skips, but also for the liquid and gaseous by-products of resource use which are more usually thought of as 'pollution'. A waste product can be solid, liquid or gaseous: its safe treatment and disposal will be highly influenced by this physical property. Gases cannot be disposed of to a landfill site, unless they have been reacted with other chemicals to render them solid. It is not generally regarded as a responsible environmental practice to dispose of solid waste (e.g. plastics) at sea or in rivers. However, many liquid and gaseous wastes are dispersed into the aquatic or atmospheric environment, sometimes after treatment, and sometimes under controlled and monitored conditions. Once discharged, they may cause pollution of air, water or soils, as may solid wastes if they decompose to produce gaseous or water-soluble by-products.

Forms of energy can also be described as wastes: waste heat from a thermal power station cooling tower, for example, will use the atmosphere and a nearby river as a sink. For noise energy from industrial processes, the surrounding locality is a sink. The word 'waste' is also used to describe environmental features that cannot currently be used for productive purposes. Derelict land rendered valueless by past industrial activity, for example, is called wasteland, as are other unproductive areas, for example the 'Arctic wastes', although this terminology is contentious as it can be argued that such wildernesses as the polar regions are valuable precisely because of their pristine and unproductive nature.

A question of balance

Wastes do not necessarily lead to environmental problems. All animals create body wastes: food and drink are consumed, useful nutrients extracted, and the residues expelled. In a sparsely populated eco-system natural processes recycle these wastes at about the same rate that they are produced. They will not build up to cause environmental problems to the producing species or others in the habitat. In contrast, the build-up of material wastes from some human activities is increasingly perceived to be causing severe environmental damage. This is evident from many of the boxes in this book, not least Box 1.1.

Box 1.1

Pesticide waste systems

It has been estimated that total global pre-harvest losses of agricultural produce to pests accounts for about 35 per cent of total production, with a further proportion being lost to vermin and fungi after harvesting. Pest control by the use of pesticides is therefore an attractive option for farmers seeking to improve their yields. An ideal pesticide would:

- attack only the target species and have no biological effects at all on others;
- give the target species no opportunity to develop strains which are resistant;
- be manufactured with no noxious by-products;
- biodegrade quickly once it had done its job.

Unfortunately, most pesticides in widespread use today do not have each of these characteristics. Some are naturally occurring compounds, prepared from insecticidal plants (for example derris, pyrethrum and nicotine). Much more common are synthetic organic compounds such as the organophosphates, carbamates and pyrethroids. These have replaced earlier synthetic insecticides which were developed and used widely throughout the world in the years following the Second World War: DDT (dichlorodiphenyltrichloroethane), lindane and dieldrin.

Increasing safety standards and the need for extensive testing of new pesticides have meant that only chemicals suitable for dealing with a broad range of pests are developed and brought to market. A pesticide which is too specific will have a small market share and may not make sufficient profit to repay its development costs. Species will differ in their sensitivity to any one pesticide, and the observed toxic effects may be very different from one species to another. This is a particular problem when insecticide use results in harmful effects on predator species, such as carnivorous insects (e.g. ladybirds and wasps) or birds. If the original target species develops resistance to the insecticide used and, at the same time, the numbers of predators are reduced due to pesticide poisoning, then pest infestations can become very difficult to manage.

Pesticides may also cause problems to species which are not part of the immediate eco-system of the target species. Because pesticides can contaminate air, ground and surface waters, as well as soils, natural air and water movements can act as dispersion media in the natural environment. Ingestion of pesticides by animals, or by mobile plants such as seaweed and plankton, can also lead to its spatial dispersion. The potential complexities are well illustrated by the story of DDT.

The insecticidal properties of DDT were first recognised in 1939 and it was initially hailed as a beneficent chemical both for crop protection and for the prevention of insect-borne diseases such as malaria. Certainly, DDT is a highly effective insecticide, in part because of its environmental persistence, which makes frequent re-spraying unnecessary. The use of synthetic pesticides was a relatively new technology in the

continued

Box 1.1 continued

1940s and 1950s and, at first, little thought was given to the environmental effects that these chemicals might have. However, the publication in 1962 of Rachel Carson's classic book *Silent Spring* triggered growing awareness of the deleterious effects of the chemical. Subsequent monitoring studies found that DDT was widely dispersed throughout global physical and eco-systems, had entered the food chain, and that its persistence had resulted in bio-accumulation in many species. For these reasons DDT was banned in the United States in 1973 and in the United Kingdom in 1984, but it is still used in some developing countries.

If the pesticide breaks down fairly rapidly after application, the opportunity for non-pest species to be damaged by the pesticide may be limited. However, some breakdown products of pesticides are themselves toxic. DDT can be metabolised to DDE, a chemical which is so persistent in the environment that its decay is difficult to measure. DDE is associated with eggshell thinning in predatory bird species, leading to widespread failure to hatch and population decline in some cases.

Some pesticides, such as the organophosphate insecticides, have effects on human health ranging from headaches and nausea to, in extreme cases, death, if exposure to large amounts of the chemical occurs. Smaller doses may have medium to long-term health effects, affecting workers exposed through manufacturing, agricultural workers, people living close to land where pesticides are applied, and those eating food which contains pesticide residues.

References: Carson (1962); Conway and Pretty (1991).

Discussion points

1 Which forms of environmental capital and environmental services were threatened by the presence of DDT in the environment?

2 This case study exemplifies problems arising from waste streams. How does pesticide use relate to other issues in Chapter 1 – resources, biodiversity, quality of life and population?

Just as the production of resources may be constrained by the ability of the environment to regenerate material and energy flows, so there may be spatial or temporal limits on the amount of waste that the environment can assimilate. For example, a landfill site has a finite spatial capacity. This can be extended by mechanically compacting the waste so that a greater mass will occupy a smaller volume. Eventually, however, technical or regulatory limits on the final height of the site will be reached and it will no longer be available for waste disposal. In many parts of the developed world there is a shortage of suitable sites to replace existing landfills as they reach their maximum capacity.

Temporal limits may apply to the ability of natural systems to dilute, disperse and degrade the waste. Carbon dioxide is a waste product from the respiration of living cells, as well as from combustion of fossil and non-fossil organic matter (Table 1.3). The carbon cycle, explained in Box 1.2, is a good example of a waste assimilation system. A stable concentration of carbon dioxide in the air will depend on the overall balance between the processes that are producing atmospheric carbon dioxide and those that are removing it. If the rate at which waste is entering such a system is greater than the rate at which it is being removed the waste will accumulate in the sink where it may cause changes in the operation of associated environmental systems.

Box 1.2

The carbon cycle and global warming

Almost all atmospheric carbon takes the form of carbon dioxide (CO_2), although a small proportion is gaseous hydro- or halo-carbons. For the level of carbon dioxide to remain constant, flows into and out of the atmosphere must be balanced. Historically (and prehistorically), there have been periods where relative stability has been achieved, interspersed with periods of rapid change. There is a striking correlation between CO_2 levels and global temperatures, colder periods, such as Ice Ages, being associated with lower CO_2 levels, and warmer periods with higher levels.

At present, the production of carbon dioxide exceeds its absorption by about 3.2 Gt each year and therefore atmospheric carbon dioxide concentrations are rising. Although carbon emissions to the atmosphere from human activities such as fossil fuel emissions, cement production and deforestation are small in comparison with natural flows through the carbon cycle it is thought that the causes of this rise, the beginnings of which coincided with the industrial revolution, are anthropogenic.

Carbon dioxide, along with methane, nitrous oxide, chlorofluorocarbons and ozone, is a greenhouse gas. This means that it interacts with infra-red radiation (heat radiation) in a way that raises, to some extent, the temperature of the Earth. Just like any other object, the Earth will remain at a constant temperature only if there is a balance between the amount of energy arriving and the amount of energy leaving. Any discrepancy in this energy balance will lead to heating or cooling. Energy arrives at the Earth from the Sun as light and heat radiation. Because the Earth is warm, it also radiates heat into space and this outgoing radiation has an average wavelength that is longer than that of the incoming radiation. Greenhouse gases have the potential to disrupt this second process by absorbing a small fraction of these longer wavelength rays so that the heat remains in the atmosphere rather than being lost to outer space. This process is called radiative forcing. Because there has been no corresponding decrease in incoming energy the overall effect of the greenhouse gases is to raise the temperature of the Earth, just as a thick duvet will raise the temperature of a sleeping child.

continued

Box 1.2 continued

Were it not for the warming effect of greenhouse gases, the average temperature of the Earth's surface would be 33°C lower than the present day figure, ~15°C, and life would not be possible. However, as a result of human activities, the atmospheric concentrations of methane, nitrous oxide, chlorofluorocarbons and ozone, as well as carbon dioxide, are rising.

The Intergovernmental Panel on Climate Change (IPCC) Working Group I is the body charged to report to the international community on the science of climate change. In its most recent report (Houghton *et al.* 2001b) it claimed there was a 'discernible human influence on global climate' and made some key points about events during the twentieth century. During the century:

- The global average surface temperature increased by about 0.6°C;
- snow and ice cover decreased;
- global average sea level rose and ocean heat content increased;
- concentrations of greenhouse gases and their radiative forcing increased;
- other forms of air pollution such as acid emissions, acted to offset, to a small extent, the effects of the greenhouse gases, by producing negative forcing;
- natural factors, such as changes in solar irradiance and volcanic eruptions, had only a minor forcing effect relative to that of the greenhouse gases.

For the twenty-first century the IPCC concludes that these changes will continue, with carbon dioxide emissions from fossil fuel burning playing the dominant causal role. A range of scenarios is modelled, taking into account uncertainties about the extent of future emissions and the ability of the environment to assimilate these. By 2100 these would result in atmospheric CO_2 concentrations of between 75 per cent and 350 per cent above the pre-industrial (1750) level of 275 parts per million (ppm). Global average temperature projections for 2100 are between +1.4°C and +5.8°C, compared with 1990. This warming will be unevenly distributed, with the northern regions of North America and northern and central Asia predicted to exceed the average significantly. Other weather changes are predicted (with varying levels of confidence) for some areas – more intense winter rainfall; dryer summers and possible drought; more variability in the duration and strength of Asian monsoons; and an increase in the severity of cyclones. Sea level rises are also predicted (see Box 1.4).

Beyond 2100 changes in climate and sea level are forecast to continue even if greenhouse gas concentrations can be stabilised, but the extent of these changes is difficult to model.

Reference: Houghton *et al.* (2001b).

Discussion point

1 Decide how global warming fits the definition of environmental problem in this chapter by defining some of the types of environmental capital, environmental service, and human needs which are important in this case study.

Waste assimilation systems

Two properties are important in understanding how wastes behave once released to the environment: dispersal and degradation. For any waste these will depend on its physical, chemical and biological properties.

For example, waste gases released to the atmosphere will often diffuse fairly rapidly away from their original source, a process aided by natural air currents. Substances which are soluble in water will also disperse, once released to the environment, through mechanisms such as diffusion and water flows, although such dispersal will be dependent on the rate at which water itself is moving. However, wastes can also accumulate. Fat-soluble substances, such as some pesticides, accumulate up the food chain as larger animals eat plants or smaller animals with the waste substance in their bodies (see Figure 1.4). If the waste substance is not broken down by the metabolism of the organisms which have ingested it, higher concentrations will be found in the predator species towards the top of the food chain than in the organisms which originally ingested the substance from the physical environment.

Bioaccumulation is a consequence of the persistence of some wastes: their tendency not to break down into other chemical forms. Other wastes do degrade and this can occur by a number of mechanisms, including bio-degradation, which occurs when organic wastes rot into simple

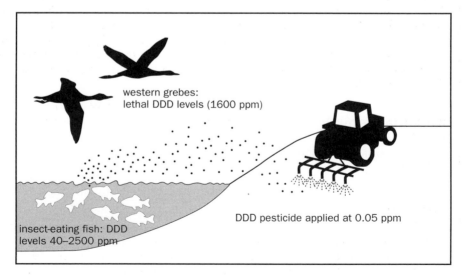

Figure 1.4 *Bio-accumulation of pesticide residues*

Source: adapted from Chesters and Konrad (1971)

chemical components due to the action of micro-organisms such as fungi and bacteria. Combustion is another common means by which wastes are degraded. When combustible waste is burned its molecular structure is broken down and new chemical products are formed. A large proportion of the mass of the waste will become gaseous (e.g. carbon dioxide, water vapour). A much smaller mass remains as ash.

Degradation is not the end of the waste story because all breakdown processes have products. Organic matter decaying in a landfill site will decompose, producing carbon dioxide, if there is oxygen present, or methane, if conditions are anaerobic (i.e. without oxygen). Ammonia is another product of bio-degradation. Such products may cause pollution if they are not controlled.

Some wastes are relatively inert: many types of plastic, such as polythene or PVC, if landfilled, will neither disperse nor degrade in the short to medium term. However, it should be clear from the above that the potential interaction of dispersion and/or concentration mechanisms and persistence and/or degradation processes within a waste system could be extremely complex. Scientific knowledge is needed to model and predict the behaviour of wastes that have dispersed into the environment and gathering the requisite data can be extremely time consuming and difficult.

Wastes and the associated problem of pollution are responsible for environmental problems at many different spatial levels, from the individual, for example asthma caused by urban air pollution, to the global (see Boxes 1.2–3). The next section examines why the production of wastes is growing each year by examining the relationship between resource utilisation, waste production and population growth.

Population growth and the environment

Population growth

During the 1990s the population of the world grew by about 1.7 per cent per year. Although this rate of growth was lower than in earlier years (in 1970 the figure was 2.0 per cent), and is expected to fall progressively, population growth is continuing into the twenty-first century. Global population reached 6 billion in 1999. The effects of the diminishing rate of growth are evident: on the lowest projection world population will stabilise by 2100 (Figure 1.5; UNFPA 2001).

Any explanation of the causes of population growth must take into account the changing patterns of global and local distribution of

Box 1.3

Sea-level rise and coastal erosion

Land and sea coexist in a dynamic relationship. Sedimentation, erosion by wind, waves and tides, and rises and falls in both sea and land levels interact to change the vertical and horizontal profile of coastlines. Until the advent of technology these processes were entirely natural but this is no longer the case. Sedimentation rates are altered in rivers subject to damming and/or canalisation. Subsiding land levels may be due to tectonic movements, volcanic activity or earthquakes or to human activity such as ground water or mineral extraction. Anthropogenic global warming of the atmosphere is thought to be increasing the severity of some weather patterns and therefore the potential erosive activity of wind and waves.

However, it is the prospect of rising sea levels, as a result of global warming, that is expected to be the most significant anthropogenic cause of the loss of coastal land to the sea in the twenty-first century. Increasing global temperatures will cause sea levels to change for two main reasons:

- As oceans become warmer, thermal expansion will lead to increases in sea volume.
- Melting terrestrial ice sheets and glaciers will increase the mass, and therefore the volume, of sea water.

Sea-level rise is likely to lag behind atmospheric warming by several decades because the transfer of heat from the atmosphere to the oceans, especially the deeper layers, is a relatively slow process.

Local responses to sea-level rise

Bird (1993) suggests that three kinds of local response are possible to the threat of inundation:

- *Evacuation and adaptation.* This is an entirely human response and does not involve management of the environment. However, there will be costs, not only to those whose land and buildings are destroyed by the sea, but possibly also to communities farther inland if their tenure is threatened by displaced coastal peoples. These costs are not entirely economic. Many coastal areas have high ecological value, for example the mangrove swamps found in coastal areas throughout the tropics and subtropical regions. Mangroves provide an important breeding ground for fish and shellfish. It is predicted that rising seas will erode the mud edges of mangrove swamps, leading to their disappearance from most exposed coastlines.
- *Maintenance of the existing coastline by engineering works.* Nearshore submerged breakwaters can reduce the erosive power of waves, especially when the tidal range is low. Sea walls and similar structures can protect land lying below high water level and the bases of cliffs subject to erosion but may also increase the rate of beach erosion seaward of the defences. Groynes are designed to slow the transfer of sand and silt down a coastline: they can be successful, but often lead to greater erosion of

continued

Box 1.3 continued

beaches elsewhere, which were formerly replenished by the drifting material. Barriers across estuaries can protect low-lying land upstream but may produce a greater risk of flooding to land seaward of the barrier. Two conditions have to be satisfied for this strategy to be implemented:

1 The threatened land must be sufficiently valuable to warrant the expenditure – this is likely to be the case only for densely populated areas.
2 There must be sufficient capital available to invest in the necessary protection works.

As sea levels rise there are likely to be many at-risk locations in the developing world where the first condition is met but the second is not.

• *Counter-attack by reclamation of intertidal and near-shore areas*. This strategy has the advantage that the high economic costs of engineering works can be offset by the value of the reclaimed land, which can be used either for urban development or for agriculture. On coasts that feature shallow bays it can be more cost-effective to barrage the entrance to the bay and reclaim the area landward of the barrage than to protect the entire coastline from erosion and flooding.

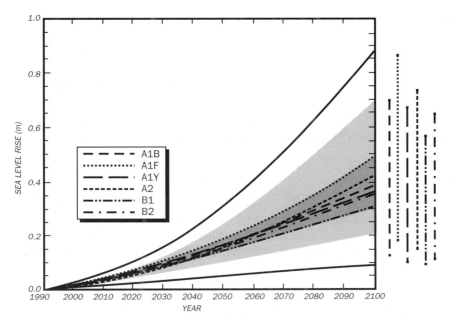

Predicted global sea-level rise, 1990–2050. Using computerised atmosphere–ocean general circulation models, thirty-five scenarios were modelled and the average prediction for each falls within the dark-grey shaded areas. The average predicted rises for six of these scenarios are identified in the key and the range in 2100 for these by the bars to the right. Light shading shows the range of all thirty-five scenarios and models. The outer lines show the increased range once uncertainties in land ice changes, permafrost changes and sediment deposition have been added in

Source: Intergovernmental Panel on Climate Change (2001), by courtesy of the Panel, Geneva

Box 1.3 continued

A global response?

One further strategy can be considered (Shennan 1992): reducing or halting the atmospheric warming which is causing sea levels to rise by reducing the emission of greenhouse gases and/or developing carbon sinks (e.g. forests) to extract carbon dioxide from the atmosphere. There are two reasons why such a strategy is of little relevance to communities threatened with sea-level rise in the next few decades. The first is that even immediate stabilisation of greenhouse gases in the atmosphere would not halt sea-level rise, owing to the time lag between atmospheric and marine warming. The second is the difficulty inherent in getting those who perceive themselves not to be greatly at risk from symptoms of global warming such as sea-level rise to reduce their carbon emissions for the benefit of those who are at risk.

References: Bird (1993); Shennan (1992).

Discussion points

How does this case study illustrate the following themes from Chapter 1?

1 The difficulty of separately identifying 'natural' and 'anthropogenic' causes of environmental changes.

2 The practical limitations of environmental management, particularly in terms of knock-on effects.

3 That changing human behaviour is in principle a more fundamental, but also a more problematic, response to environmental problems than environmental management?

population. In 1990 the world population of 5.3 billion was distributed between the developed and less developed nations in the ratio of approximately 1:3. Because population is growing more rapidly in developing countries, by 2025 this ratio is expected to be closer to 1:5 (World Bank 2002).

The fundamental cause of population growth is falling death rates, due to improvements in public health: for example, vaccines, malaria control methods and reductions in child mortality. Advances in agriculture and food distribution have reduced the chance of dying of famine for inhabitants of some countries. As more children survive to reach reproductive maturity the number of births increases. The number of children born to each couple is obviously important in determining the rate of this growth. In most countries of the world the total fertility rate (TFR) of live births per woman is falling, but only when the average

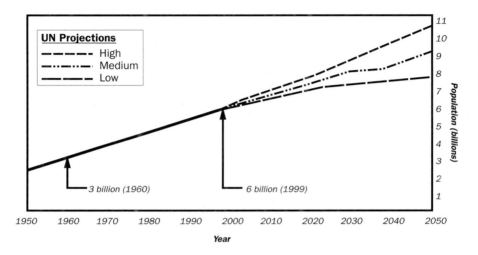

Figure 1.5 *World population growth, actual and projected, 1950–2050*

Source: UNFPA (1999: 3), updated UNFPA (2002), by courtesy of the UN Population Fund, New York

number of births per couple reaches the replacement level (two) is it possible for population stability to be in prospect. (See Table 1.4.) The use of contraception is vital to achieving population stability in any region.

Population, resources and waste

These patterns of population growth have implications for the demand for resources, the rate at which wastes are produced, and the operation of waste systems. It is important to recognise that 90 per cent of the global population increase is taking place in the less developed countries, where the per capita demand for resources (and therefore the production of wastes) is currently much less than in the developed world. For example, the industrial emission of carbon dioxide (an indicator of fossil fuel use as well as atmospheric pollution) was 19.8 t per capita in the United States and 1.1 t per capita in India in 1998 (World Bank 2002).

However, it is not just population growth that increases the rate of extraction of resources and the production of wastes. As material living standards rise, rates of production and consumption increase in a way that compounds the effect of population growth on resource throughput. It is estimated that the population of India was increasing by 1.8 per cent per annum at the turn of the millennium; economic growth, however, was occurring at the higher rate of 3.9 per cent, meaning an average increase in material wealth both in absolute terms and per capita (World Bank 2002).

Table 1.4 Trends in total fertility rate and population growth by country and region

Country	Total fertility rate		Population growth (%)		Total population, 2000 (million)
	1997	2000	1997	2000	
World	2.77	2.68	1.43	1.31	6,058
China	1.90	1.90	1.02	0.90	1,262
East Asia and Pacific	2.17	2.12	1.20	0.94	1,809
Europe and central Asia	1.67	1.58	0.13	0.08	474
India	3.30	3.06	1.76	1.83	1,016
Latin America and Caribbean	2.69	2.56	1.57	1.52	516
Middle East and North Africa	3.69	3.42	2.05	1.90	295
South Asia	3.52	3.29	1.89	1.92	1,355
Sub-Saharan Africa	5.47	5.20	2.82	2.43	659
United States	2.03	2.13	1.25	1.19	282

Source: World Bank (2002)

In countries that are making the transition from an agricultural to an industrial economy, most of this population growth is taking place in urban areas. The demographic factors described above are compounded by migration into cities by rural dwellers seeking an improved standard of living. For example, the World Bank predicts that the number of people living in African cities will rise from 297 million in 2000 to 766 million by 2030 (World Bank 2002). One effect on the increasing urbanisation of the global population is to exacerbate environmental problems associated with waste systems. Spatial concentration of the production of human, municipal and industrial wastes can overload waste systems, leading to the accumulation of waste products and resulting in air, water and land pollution. This in turn can diminish environmental capital by reducing or destroying the ability of the environment to provide environmental services.

The key point is the potential of wastes to disrupt environmental and biological systems. *Homo sapiens* adapted, through evolution, to fill an environmental niche shaped by the interlinked systems of climate, atmosphere, the hydrological cycle, the geosphere and life itself, and therefore the human species has a vested interest in the stability of these systems. Despite this there is increasing evidence (for example, global warming and the acidification of fresh waters and soils from pollution caused by fossil fuel burning) of serious disruption to them due to the accumulation of wastes and their by-products.

Biodiversity

The basis of biodiversity is the genetic variation between organisms and species. At the heart of almost every living cell is a nucleus that contains the genetic material that the organism (animal, plant or micro-organism) has inherited from its parents. This material consists of long chains of deoxyribonucleic acid bases (DNA). There are four of these bases (adenine, thiamine, cytosine and guanine) and they can be arranged in an infinite number of ways, just as the ten digits in the Arabic counting system (0, 1, 2, 3, 4, 5, 6, 7, 8, 9) can be used to express an infinite range of numbers.

A gene is a DNA sequence that provides the template from which a particular protein can be manufactured in a cell from its component amino acids. In this way DNA provides the blueprint for the organism, specifying all its inherited characteristics. Some of these characteristics will be found in all members of the species: all rats, for example, have four legs, fur, sharp teeth, whiskers and a tail. Other characteristics will vary between individuals within a species: eye and fur colour, for example.

The variation in DNA sequences between species and within species is known as genetic biodiversity. It has arisen as part of the process of evolution. When cells split to form new cells, the DNA in the nucleus is copied so that each new cell will have the same genetic material as its parent. Occasionally a mistake may occur in the copying process so that the new DNA is not an exact copy of the old. This mistake may mean that the new cell is unable to synthesise a vital protein, as the coding instructions are now wrong: in that case the cell will die. Alternatively, the change may not affect the functioning of the cell very much, or may even improve it in some way. A non-fatal change to a reproductive cell can lead to this change being passed on to future generations.

Over generations these genetic changes are compounded and species evolve. The genetic changes interact with changes in the environment of the species (food, habitat, climate, predators etc.) in the process of evolution by natural selection. Individuals which are genetically most suited to their environment are most likely to survive to reproduce, passing on their genes to their offspring. As 'successful' genes spread through a species, less successful genes will be progressively eliminated from the gene pool. Over very long periods of time, two populations of the same species, unable to interbreed and share their gene pool because of geographical isolation, may evolve into two completely different species.

Genetic diversity is therefore the biological basis of two other forms of biodiversity: species diversity and eco-system diversity. Species diversity is simply the range of species currently living in a particular geographical area. Eco-system diversity refers to the range of species (and sometimes the genetic diversity within these species) participating in a particular eco-system or set of eco-systems. Although there is agreement that a richer range of genetic make-ups or species is equivalent to greater biodiversity, measuring biodiversity is a complex and difficult undertaking (Gaston 1996: 3–5).

Table 1.5 summarises the arguments for preserving species biodiversity. To the extent that biodiversity is valuable to humankind, it can be looked on as environmental capital and as a non-renewable resource. The development of biodiversity through evolution is a very slow process – much slower than the rate at which this diversity is at present declining. The decline is as a result of both the overall reduction in the number of species and the loss of genetic diversity within species, due to selective breeding of some food crops and population decline of other species.

Table 1.5 *Arguments for conserving species*

Argument	Rejoinder
Humankind has moral and ethical responsibilities to care for life on earth	This argument cuts little ice with barbarians
Species such as flowers and butterflies enrich our lives	Many species are not attractive (e.g. slugs). Also, it is difficult to put a money value on species (see Chapter 8)
Species can turn out to be valuable e.g. as pharmaceuticals	Most will never be economically useful and (but) we have no way of identifying those which may turn out to be so.
Species have functions within the systems which form part of the earth's life support systems, which will be weakened by biodiversity loss	Possibly, but this is not scientifically proven
If humankind is unable to maintain biodiversity, we are also unlikely to achieve civilised human activities and a decent quality of life	Sceptics who reject the above arguments are unlikely to be convinced by this one

Source: based on Kunin and Lawton (1996)

These losses may be due to direct hunting, collection or persecution of animals and plants by humans, or are more likely to be the result of damage to, or destruction of, species habitats. The growth in throughput through the resource cycle puts pressure on habitats in many different ways. Increasing demand for mineral, forestry and agricultural products is fuelling rapid changes in land use patterns, compounding the effects of urbanisation and industrialisation. Wastes, and the knock-on effects of their accumulation on the environment, such as global warming, are causing the deterioration of habitats and biodiversity throughout the world. Lack of scientific data makes the absolute rate of loss impossible to compute with any accuracy, although it has been estimated that up to 50 per cent of the world's plant species may be at risk (Pitman and Jørgensen 2002). Global species losses of animals, plants and micro-organisms through extinction have been estimated to be occurring at between anything between 1 per cent and 30 per cent per decade (Stork 1997: 62–3).

Quality of life

Quality of life and environmental capital

Individuals do not just need the environment for the resources needed to provide them with consumer goods:

> The family which takes its mauve and cerise, air conditioned, power-steered and power braked automobile out for a tour passes through cities that are badly paved, made hideous by litter, blighted buildings, billboards and posts for wires that should long since have been put underground. They pass on into a countryside that has been rendered largely invisible by commercial art. . . . They picnic on exquisitely packaged food from a portable icebox by a polluted stream and go on to spend the night at a park which is a menace to public health and morals. Just before dozing off on an air mattress, beneath a nylon tent, amid the stench of decaying refuse, they may reflect vaguely on the curious unevenness of their blessings.
>
> (Galbraith 1987: 208–9)

A high standard of living, measured in terms of personal and individual access to resources and products derived from resources, is not necessarily an indicator of a high quality of life. This latter concept takes into account a much wider range of factors than merely material goods and environmental capital is fundamental to these.

Aesthetic enjoyment of the natural environment is an important component of quality of life for many people and thus the ability of the environment to provide such enjoyment can be conceived as an example of environmental capital providing an environmental service. This ability can be adversely affected, as the quotation above demonstrates, by the accumulation of wastes, or the diminution of environmental value by, for example, landscape despoliation by resource extraction. Enjoyment may involve active recreation such as hill walking in unspoiled areas of wilderness, or just opening the window for some fresh air. It is sometimes vicarious: some people who know that they will never visit Antarctica, or a tropical rain forest, still care passionately about these areas and will derive pleasure from a knowledge of their continued existence and ecological health. Indeed, some would rank the enjoyment of a healthy and undefiled environment as a spiritual necessity, not a mere aesthetic enjoyment.

Safety and security can also be thought of as environmental services that contribute to quality of life. Those who know they are at risk from natural hazards, such as floods, earthquakes or volcanoes, are likely to experience anxiety and stress, even if the actual event they worry about never actually comes to pass. Other environmental hazards may result from human behaviour and may be therefore more susceptible to human management to reduce or remove the risk. The streets of medieval cities were a health hazard owing to the lack of effective systems to deal with human and other wastes from concentrated population clusters – as is still the case in many cities in the developing world today. Huge investments in public health engineering mean that these days, at least in the developed world, raw sewage is unlikely to be encountered by the average city dweller and the resulting epidemics of cholera, dysentery and bubonic plague are beyond living memory.

But industrialisation has brought new environmental hazards to be faced. Some aspects of the industrialised environment may adversely affect human health, causing chronic or acute disease. Traffic accidents, excess noise, poor air quality, lead in drinking water, pesticide residues in food, as well as the stress which can derive from the pressures of supporting an affluent lifestyle or enduring a poverty bound existence – all these factors cause real or perceived risks to health and can therefore detract from the quality of life. These hazards are mostly encountered at a local scale, but increasingly global environmental problems, such as global warming or stratospheric ozone depletion, are beginning to be perceived as serious risks to human health and well-being.

Quality of life and social capital

Quality of life also depends on factors that are to do with the social, rather than the natural, environment. Social capital can be defined as the formal and informal structures and mechanisms within society that produce a stream of benefits for individuals and communities. Examples of the types of benefits that social capital can provide, and are important to quality of life, might be job security; social security; freedom from racial or sexual harassment; freedom from the fear and the actuality of crime; and access to high quality services for health care and transport.

Environmental problems

The preceding section identified resources, wastes, population, biodiversity and quality of life as environmental 'issues': that is, areas where the provision of environmental services to meet human needs is problematic. Environmental 'problems', and more importantly the search for policy-based solutions, are, of course, the subject matter of this book so some consideration is required of what is meant by the phrase 'environmental problem'. The proposed working definition of environmental problem is: *an environmental problem can be said to have arisen when the provision of environmental services is insufficient, either in quantity or in quality, to meet human needs.*

It is important to note two important characteristics of this definition:

- Interpretations of the words 'insufficient' and 'human needs' are subjective. Therefore claims about the existence and significance of environmental problems are also subjective and often contentious.
- Defining environmental problems in terms of an insufficiency of environmental services does not necessarily imply that the root causes of the problem are environmental. Human over-use or abuse of these, or other, environmental services might be to blame.

The final sections of this chapter will expand upon these points whilst introducing the more detailed discussions which are to be found in subsequent chapters.

Problem? Whose problem?

As the discussion of population growth earlier in this chapter demonstrated, the Earth is supporting record numbers of human beings. Many environmental services are at present unproblematic, for example the basic human need for oxygen. As long as there is sunlight, and there are sufficient green plants, terrestrial and marine, to convert carbon dioxide to oxygen through the process of photosynthesis, the Earth's atmosphere will continue to be composed of approximately one-fifth oxygen.

Other environmental services are problematic in some circumstances. A proportion of the human population suffers from poverty; malnutrition; poor air and water quality; overcrowding; or pollution in both the developed and the developing world. Inevitably, some people are more badly affected than others. The significance of this unequal distribution of the effects of environmental problems will be explored in more detail in Chapter 2, but for now it is sufficient to note that perceptions of environmental problems, and support for proposed solutions, will differ depending on one's situation. A proposed landfill site will cause great concern to those living within a 5 km radius, who might fight their corner at a public inquiry by advocating alternative sites or alternative strategies of waste management. Those living 50 km away may not be concerned much as long as they continue to receive a cost-effective waste management service. Thus one person's environmental problem can be another's environmental solution.

Relevance is also an issue, especially when the human needs that are being denied are aesthetic or vicarious. The point of view of those owners and workers who will profit from a proposed quarry need to be balanced with that of residents who value their views of the unspoiled landscape. Poor peasant farmers in South America or Indonesia will have difficulty in accommodating the vicarious 'need' of European environmentalists, who may never even visit their country, for virgin rain forest to be protected.

But differences of opinion about the existence and severity of environmental problems do not just depend on whether perceived interests are advanced or damaged in any particular set of circumstances. Consider the following two quotations on the prospects for humankind overcoming the major environmental problems introduced earlier in this chapter.

> In short, while there are ecological rules on this planet – naturally – there are no obvious limits that we know of. Warmth, movement, food, shelter, education and all the other things we prize as humans can all be provided in dozens of different ways. Our task now is to find new ways of achieving them. That has always been our task. If this resource, or that pollution sink, or that production method has proved redundant or unsustainable, we must merely find another or find ways of making fewer demands.
>
> (North 1995: 281)

> Humanity, this sudden new evolutionary development which has had such a high impact upon the planet, now threatens not only its own survival but that of many parts of the biosphere itself. The rapidly manifesting global crises, the long shadow, cast by the fast-growing figure of humankind, is stretching into the very heart of our biosphere. Since wandering early tribes began to fire the forests, this shadow has spread across land and ocean, through air, water, and soil, into space and deep into the life-blood of evolution itself.
>
> We might look upon our global crises as a challenge as well as a threat – we are sitting our final evolutionary examination for our viability as a species. Unfortunately, the time limit is rapidly approaching.
>
> (Myers 1994: 17)

Whilst both authors acknowledge the existence of some serious environmental problems they differ markedly in the conclusions they draw about significance of these. In part this is due to the difficulties of gathering and interpreting scientific information about the environment, which are discussed in Chapter 4. But differences of opinion about the importance of environmental problems persist even in circumstances where the scientific facts are fairly certain. The evidence may be objective but its interpretation is always subjective and related to human needs for environmental services. There is more concern about the extinction of rare orchids, which are valued for aesthetic reasons, than there is about obscure grasses, mosses or fungi. Long-standing campaigns to save the whale have been tremendously successful, but less attractive marine species, such as plankton, enjoy minimal public concern.

Of course, the very concept of environmental services is inherently anthropocentric as it focuses only on aspects of the environment that are important to human welfare. Anthropocentrism is also inevitable in the definition of environmental problems and the above definition makes this explicit. Including aesthetic and spiritual human needs widens the scope beyond the purely material but does not alter the anthropocentric nature of the definition. Nothing could. 'Problems' are always subjective: human beings can interpret environmental information and define their need for environmental services (and therefore the existence and significance of environmental problems) only from their own perspective. Although this perspective can be individual or collective, humans cannot adopt the perspective of other species. Conflicts often arise when the same objective information is interpreted differently by groups with different priorities and value systems. The role of values in the formation of environmental attitudes is discussed in more detail in the next chapter.

Environmental problems or human problems?

Environmental problems, then, are always human problems in the sense that their effects are the denial, to some extent, of human needs. But what about the causes of environmental problems? Do they lie within the environment, as the phrase might suggest, or are these in fact human problems in terms of their causes as well as their effects? For any given environmental problem the answer is usually more complex than the question suggests, as Box 1.3 which examines coastal erosion and sea level rise, exemplifies.

One class of environmental problems consists of those due to natural occurrences such as earthquakes, floods and droughts. The eruption of the volcano Vesuvius which caused such severe environmental problems for the inhabitants of Pompeii and neighbouring settlements in AD 79 that

2000 people died was entirely environmental in origin. However, should Vesuvius stage an eruption in the present day, as vulcanologists predict is likely, the devastation and death toll would be far greater than it was in Roman times and the cause of this catastrophe would not lie entirely with the environment. In the light of twentieth-century scientific knowledge, the dense development which has sprouted around the Bay of Naples, coupled with a poor transport infrastructure which would greatly hinder evacuation of the area, appears extremely unwise (Oak Ridge National Laboratory 2002). Similar comments could apply to urban developments in earthquake zones, such as the southern California conurbation. Thus, the potential consequences of environmental problems caused by natural hazards, such as volcanoes and earthquakes, can be exacerbated by human behaviour.

Other problems are produced mainly or entirely as a result of human activity. The pesticide case study from Box 1.1 and several other examples from later chapters (e.g. Boxes 2.4, 4.1, 4.4, 5.1, 7.1, 7.3) are wholly anthropogenic. In the case of global warming, unpicking the natural from the anthropogenic is extremely difficult, but the scientific consensus is that there is now evidence of a human influence on global climate (Box 1.2).

So it can be seen that in the case of both natural and anthropogenic hazards environmental problems usually result from the interaction of environmental and human factors. Although it is not always possible to distinguish between the effects of the natural and the anthropogenic it is often the case that human factors are more significant, for reasons that are introduced in the next chapter.

Environmental policy and environmental problems

So much for problems, what about solutions? Environmental policy is the key to avoiding, solving or ameliorating environmental problems. The working definition of environmental policy used in this book was set out in the Introduction: *environmental policy is a set of principles and intentions used to guide decision making about human management of environmental capital and environmental services.*

Most of the case study boxes in this book show that, for environmental problems wholly or partially caused by human behaviour, changing this behaviour is usually the most fundamental approach to environmental problem solving. But managing the environment in an attempt to solve an environmental problem – by, for example, building coastal defences – is itself a form of human behaviour. So, the purpose of environmental policy is to change human behaviour – to make people act in ways which do not generate environmental problems, or which generate problems of lesser significance than was previously the case.

It is, of course, far easier thus to describe the general aims of environmental policy than it is to design and implement a successful version of the same in respect of a particular environmental problem. The scale of the task for environmental policy makers is enormous, given the trends in the depletion of environmental capital described in the first half of this chapter. The potential rewards, in terms of meeting the needs for environmental services of the increasing numbers of people on the planet, make the endeavour imperative. Subsequent chapters aim to provide a guide to successful environmental policy making, including the potential pitfalls and uncertainties. The first step to changing human behaviour is understanding it, so Chapter 2 examines why people behave in ways that cause environmental problems.

Further reading

Follow-up reading for this chapter falls into two categories: texts that take a broad approach to environmental problems and those that examine one or more in greater detail. In the former category are student textbooks written from a mainly scientific perspective such as Goudie (2000), Park (2001) and Miller (2001). Brown (2001) gives an up to date assessment of the state of the planet and some ideas for saving it but is marred by a simplistic approach to the politics of doing so, especially when considering the relationship between poverty and population growth. A very different analysis is offered by Lomborg (2001), who argues that environmental groups have exaggerated both the scale and the intractability of environmental problems, a view shared by North (1995). Books on specific topics include the following. *Resources*: McLaren *et al.* (1998); Redclift (1996); Deffeyes (2001). *Waste and pollution*: Harrison (2001); Pepper *et al.* (1996). *Population*: Leisinger *et al.* (2002). *Biodiversity*: Wilson (2002).

2 The roots of environmental problems

The causes of over-use of environmental capital by humans:
- **The biological nature of the human species**
- **The nature of human needs and wants**
- **The conflict between individual and larger scale interests**
- **The role of values in determining attitudes and behaviour**
- **The extent to which people are prepared to take into consideration the long term consequences of their actions**

Introduction

Identifying human behaviour as responsible, to a large extent, for the formation of environmental problems does not explain the reasons for such behaviour. These underlying reasons must be understood if such behaviour is to be changed. The five broad themes identified above are examined in turn in this chapter to try to identify the human roots of environmental problems.

Human nature

What makes humans special?

Homo sapiens is unique among living species. As the human brain has evolved, three areas within it in particular have developed to a greater extent than in any other primate (Bronowski 1973):

- the area which controls the hand (giving greater dexterity);
- the area which controls speech and language;
- the frontal lobes which, amongst other things, allow us to imagine future events.

Our greatly enhanced skills in these three areas have two important consequences for our relationship with our environment. Dexterity, combined with our ability to communicate ideas and information, has allowed the development of technology, exponential population rises and increases in the usage of environmental services, with the actual and potential consequences which were discussed in Chapter 1. Resultant environmental problems, such as pollution, have existed at least since the development of the most fundamental energy technology, the ability to create and use fire. Archaeological evidence suggests that indoor smoke pollution has been troublesome to humankind for thousands of years (Brimblecombe 1987).

The second consequence springs from the combination of the abilities to communicate and to envisage the future. Although the effect of some animal behaviour is to provide for the immediate or longer-term future (by building nests in which to lay eggs, or by laying food by in the autumn), it is generally accepted that this is a result of instinct, not intellect. In contrast, human awareness and knowledge about environmental problems can be shared and developed, and past events used to form the basis of future predictions. Visions of the future can be used to justify the deferment of gratification: putting aside the possibility of present benefits (or accruing present costs) in order to secure greater gains at a later date. Thus, the human capacity to create environmental problems is matched by a capacity to plan and to carry through ameliorative solutions.

Nature and nurture

But to what extent is it possible to change human behaviour, through environmental policy or by any other means? This question cannot be answered without reference to debates about the extent to which human beings are programmed by their genetic make-up, and how much by their upbringing and other cultural influences. Socio-biology is a school of thought developed in the 1970s (Wilson 1975) which holds that there is a strong genetic basis for human behaviour, whilst acknowledging that cultural factors also play a part. Socio-biology has sought to explain both negative traits such as aggression, genocide, xenophobia and territoriality as well as positive characteristics like altruism, love and ethics. It does this by applying theories of natural selection not to the individual, but to genes shared by individuals within groups. This is the theory of the selfish gene (Dawkins 1976).

In any group of animals sharing common ancestry certain genes will be common. All the behavioural traits listed above can contribute to the survival and reproduction of the group as a whole, even if it is at the expense of the individual. Therefore, according to socio-biological theory, such traits will be favoured by natural selection. Groups with a gene pool giving them the potential for such behaviour have an evolutionary advantage over those which do not. These genes are likely to be successfully propagated. This is not to say that the relationship between genes and behaviour is simple. No one suggests that there are individual gene codings for particular behaviours: there is no 'gene for aggression', for example, nor a 'gene for altruism'. The mechanisms by which genes determine behaviour are complex and poorly understood, but anyway it is generally agreed that genetic factors alone cannot provide a full explanation.

The argument over the extent to which human behaviour is predetermined by genetics, determined by the familial and social environment into which people are born, or is the result of freely made choices made by individuals, is one which has preoccupied philosophers and politicians throughout history. All but the most extreme analyses acknowledge that all three factors, in fact, play a part. The key area of debate concerns the relative importance of the three.

Critics of socio-biology (e.g. Rose 1984) point out that complex systems, such as human beings and human cultures, cannot be explained solely in terms of their individual components. To attempt to do so is reductionist, a concept discussed further in Chapter 4. The whole is greater than the sum of its component parts, and genetics can no more be used as a complete explanation for behaviour than can sub-atomic physics wholly to explain genetics. According to Rose, psychology and sociology are the disciplines most likely to produce useful explanations of human behaviour. Disciplines dealing with lower levels of organisation (physiology, biochemistry, genetics, molecular biology, sub-atomic physics) can sometimes provide insights which complement or enhance the higher scale explanations but can never provide the whole story.

Some of the behavioural traits which have been scrutinised by the socio-biologists – for example altruism, ethics and xenophobia – are significant in the avoidance of, or causation of, environmental problems. If human behaviour is largely predetermined, whether by genetic or by cultural factors, the task of the environmental policy maker trying to change that behaviour becomes correspondingly harder.

Human needs and environmental capital

What do people 'need'?

In his book *Motivation and Personality* the psychologist Maslow (1970) proposed that human needs can be arranged in a hierarchy, with the most fundamental at the bottom (Figure 2.1). Only when lower-level needs are being satisfied, Maslow suggests, will individuals be able to attend to higher-level needs. The most fundamental needs, of course, are those concerned with staying alive, so the two lowest levels of the hierarchy address these. First there are the physiological needs for food and water which form the base of the hierarchy. At the next level up are needs concerned with safety – for example, shelter and security.

Maslow suggests that it is only when individuals are fed, watered and safe that they can devote their attention to meeting the higher level, non-material needs. At the third level is the need for loving relationships and a sense of belonging with a social group. When this is secured the need to be esteemed by one's fellows can be worked for. The highest level needs in the hierarchy are those of self-actualisation – intellectual, spiritual and aesthetic fulfilment.

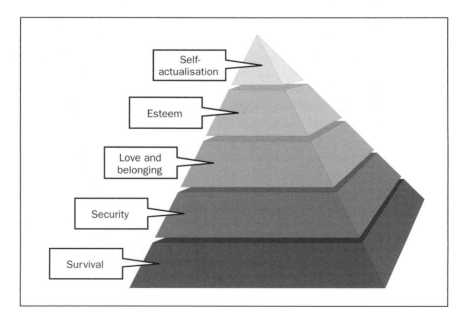

Figure 2.1 *Maslow's hierarchy of needs*

Maslow's hierarchy cannot be applied in a sweeping and rigid way to any individual or situation. For example, it would be nonsense to claim, purely on the basis of the hierarchy, that homeless people are spiritually underdeveloped. Like many other models developed in social science the hierarchy of needs is ideal-typical. This means that its use is as an aid to the interpretation of human behaviour, in particular by allowing comparisons between the ways in which different groups of individuals in differing circumstances meet their needs in different ways.

Meeting human needs at the lowest hierarchy level has the obvious potential to create environmental problems as food is an environmental service (see Boxes 1.1, 4.1–2 and 7.2). Ten thousand years ago the first agricultural societies emerged: previously human tribes had sustained themselves through a hunting-gathering existence. The shift to agriculture meant that more food could be produced per unit of land, thus offering the possibility of a better material standard of living. There was also the enhanced security of food supply which would have seemed especially valuable to people who had previously depended on the vagaries and fluctuating population of the local wildlife for their survival. The links with the lower levels of the hierarchy are clear and hence the motivation for this shift. Agriculture implied a settled way of life and was the essential precursor of the development of early cities such as Ur in Mesopotamia. Once established, such cities tended to grow rapidly as the increased food supply enabled population growth: indeed, it has been suggested that it was the need to accommodate a growing population that, over a long period of time, gradually forced humans to adopt the agricultural way of life, which is much more labour-intensive than hunting-gathering (Roberts 1998).

By the standards of the twentieth century such civilisations enjoyed a very rudimentary standard of living. However, it was not long after their establishment that some of these early settlements encountered problems associated with the deterioration of their local environment (Ponting 1991: 68–87). Even subsistence standards of living can impose strains on the environment if inappropriate management is practised or if the population load is too great. Affluence is clearly a driver of the over-use of environmental services. But poverty does not guarantee limited environmental impact – rather the reverse. Those following subsistence lifestyles will have little choice but to over-exploit the environmental capital they manage if the alternative is famine.

Needs and satisfiers

Once needs at a higher level in the hierarchy are being met the potential for over-use of environmental services becomes greater. On the face of it, this is surprising. Maslow's higher-level needs are fundamentally emotional, not material: it is not necessarily the case that the needs for love, esteem and self-actualisation will involve over-consumption of environmental capital. However, such needs are, in fact, commonly met in ways which require the accumulation of material goods. Max-Neef (1991) usefully distinguishes between needs, on the one hand, and satisfiers on the other. The need for subsistence, for example, is met, in part, by food and water, both of which are satisfiers. Esteem needs can be met in a multitude of ways, for example through consumption (keeping up with the Joneses) or through other activities or attributes (charitable giving, participation in the community, an immaculate front garden).

> The way in which needs are expressed through satisfiers varies according to historical period and culture. The social and economic relations, defined by historical and cultural circumstances, are concerned with both the subjective and the objective. Hence, *satisfiers are what render needs historical and cultural, and economic goods are their material manifestation.*
>
> (Max-Neef 1991: 27, emphasis in original)

Satisfying one set of needs can interfere with or prevent the satisfaction of another set and in extreme cases meeting higher level needs can take precedence over even physiological needs, as the case study in Box 2.1 demonstrates.

Box 2.1

Easter Island: statues and status

Easter Island had extensive tree cover when it was originally settled by Polynesian migrants in the fifth century. Trees were cut for fuel, house construction and canoes, which were used for fishing. Chickens were kept and sweet potatoes cultivated. Maslow's lower-level needs of food and shelter were thus easily met, leaving islanders with plenty of spare time for activities directed at meeting higher-level needs.

Archaeological evidence seems to suggest that, as the culture developed, clans dominated by powerful élites competed with each other through the construction of ever more of the massive statues for which the island is most famous. The statues seem to have had a religious, as well as a cultural, significance. These were erected at coastal

continued

Box 2.1 continued

sites but fashioned at inland quarries. As there were no horses or oxen, human power was used to drag the statues across the island on trackways made from tree trunks.

Easter Island statues
Source: photo by courtesy of Rachel Bridgeman

The entire society seems to have collapsed in the mid-sixteenth century, just after a period when statue construction had escalated rapidly. Ponting identifies the cause of the collapse as the almost total deforestation of the island, and the most important factor driving this process as the felling of trees for use as trackways. Yet statue transportation continued up to and beyond the point where the lower-level needs of the population were threatened by lack of wood. Of the 600 statues found on the island, over half were under construction inland, presumably stranded through lack of rollers. With the woodlands destroyed, caves, stone shelters and reed huts became the only means of shelter. Timber canoes were replaced with inferior reed boats, so fish yields declined. Soil erosion due to lack of tree cover damaged sweet potato production.

'Against great odds the islanders painstakingly constructed, over many centuries, one of the most advanced societies of its type in the world. For a thousand years they sustained a way of life in accordance with an elaborate set of social and religious customs that allowed them not only to survive but to flourish. It was in many ways a triumph of human ingenuity and an apparent victory over a difficult environment. But in the end the increasing number and cultural ambitions of the islanders proved too great for the limited resources available to them. When the environment was ruined by the pressure, the society very quickly collapsed with it, leading to a state of near barbarism' (Ponting 1991: 6–7.)

This story reinforces the point made in Chapter 2 that the hierarchy of needs cannot be applied in a rigid way. The islanders, collectively, must have become aware well before

Box 2.1 continued

the collapse that future food production was threatened by the deforestation caused by statue transportation. Yet they did not have the ability, as a society, to stop – to sacrifice the ambitions of the élite islanders in order to secure subsistence for themselves and their descendants.

Reference: Ponting (1991), chapter 1.

Discussion points

1   How can Maslow's hierarchy be used to account for the motivation of the islanders?

2   Imagine archaeologists in ten thousand years time are studying the early twenty-first century. Which current consumer goods might puzzle them in the way that the Easter Island statues puzzle contemporary scholars?

Needs in a postmodern society

In recent decades there have been massive changes in the production of goods and services which have been accompanied by equally large changes in consumption patterns. These changes have been characterised by some analysts as components of the shift from a 'modern' to a 'postmodern' society. Other cultural changes associated with this shift will be introduced in Chapter 4 when analysing the relationship between science and society.

When used in this context, the term 'modernity' refers to:

> the social order that emerged following the Enlightenment. Though its roots may be traced further back, the modern world is marked by its unprecedented dynamism, its dismissal or marginalization of tradition, and by its global consequences. Modernity's forward-looking thrust relates strongly to belief in progress and the power of human reason to produce freedom. But its discontents spring from the same source; unrealized optimism and the inherent doubt fostered by post-traditional thought.
>
> (Lyon 1999: 25)

The industrial revolution, the rise of the firm organised on the basis of bureaucracy and the production line, and the resultant production of an abundance of material goods were all manifestations of the modern age. Work is of course a satisfier. It meets human needs for subsistence indirectly (through the products of labour, either food, shelter or money).

It can directly meet needs for belongingness, esteem and self-actualisation also, but some types of work are more likely to do this than others. In the modern age production methods involving assembly lines and the fragmentation of tasks so that workers have no vision of the overall process, and therefore no identification with the overall product, have been termed 'Fordist', after Henry Ford, who pioneered these methods in early motor car production. For the workers the result was loss of the ability of work to meet higher-level needs and, instead, a sense of alienation from the work process (Braverman 1974).

Postmodern production is characterised by the new forms of work, in information, communication and travel based service industries, which have arisen to replace the industrial jobs which vanished from European and North American economies in the 1980s. Postmodern analyses point to the fact that the distribution of work has also changed. Large companies now tend to employ a core of key staff, out-sourcing semi-skilled and unskilled work to temporary workers or smaller companies, sometimes on the other side of the world. Pressure for constantly increasing productivity and flexibility of production, both in the public sector (much diminished in size by privatisation programmes) and in the private sector, lead to stressful working lives.

But production is only half the equation. Emerging and established trends in consumption patterns and behaviours are central to postmodern analyses (Lyon 1991). Shopping has become a leisure pursuit, to the extent that some supermarkets now open twenty-four hours a day. An unprecedented variety of goods is available to cater for each and every taste and desire. The advertising industry has grown in economic and cultural importance – its role the production of 'needs' in television viewers, magazine readers, Internet surfers or those walking or driving down the street past poster hoardings. Style and fashion are a predominant preoccupation, even for young children, so that the logo on a pair of running shoes can justify a price tag several times higher than that on the unbranded version. Increasingly the image associated with the consumed object becomes more important than the object itself. Goods which are still serviceable may be discarded because they are no longer fashionable or because a newer version is desired.

The work of Maslow and Max-Neef is important to an understanding of postmodern consumer culture. For example, subsistence needs are met (for those who have access to sufficient funds) by food and drink purchased from the supermarket. But even basic goods like these can act as satisfiers for higher level needs. The bottle of cola may have been chosen less for its taste than for its image, as presented by advertising, as the drink of choice of attractive and popular teenagers, and is therefore acting as a satisfier at the love and belongingness level. Products perceived as exotic and sophisticated ('designer water', fruit and vegetables from the other side of the world) may be simultaneously a response to needs at the esteem level (I am as sophisticated as my neighbours), the cognitive level (I am curious about Thai food), the aesthetic level (I enjoy fine wine) or even at the self-actualisation level (by realising my desires I become a better person).

Of course, satisfiers purchased with the intention of meeting a particular need may not necessarily satisfy. It might be thought that a lonely teenager is likely to find no more than temporary comfort from the bottle of cola: friendship is what s/he really requires. Max-Neef (1991: 31–2) classifies such goods as 'pseudo-satisfiers': they are acquired in the hope of meeting a need but can only give 'false satisfaction'. This approach can be criticised, however, as paternalistic. An important feature of postmodern consumer culture is the scope it gives for individuals to express their preferences in highly individualistic ways. Such choices are inherently subjective and so are any judgements as to the authenticity or falsity of the satisfaction gained. What is clear is that wasteful patterns of consumption in the developed world are damaging environmental capital and therefore, potentially, the future availability of environmental services. The parallel with Easter Island (Box 2.1) is uncomfortable.

The 'tragedy of the commons': models and morals

Of course, those who consume goods as satisfiers for their needs are not necessarily the same people who are adversely affected by the environmental problems consequent upon the production, consumption, operation and disposal of these goods. A common problem for environmental policy makers is the situation where the satisfaction of the needs of one group is at the expense of the wider community. One well known way of conceptualising this situation is through the use of the tragedy of the commons model. This analogy was developed in the context of global population growth, and related problems such as pollution, in 1968 by Hardin. The basic message is that environmental capital to which there is open access will inevitably be over-used.

The model proposes a community of householders, each owning one cow. Each has access for grazing to a piece of common land. At the outset the number of cattle grazed is less than the maximum which the land can support. However, because the cows are privately owned, but the common is public, each household has an incentive to increase its holding of cattle if it can. The additional milk and meat from each extra cow will belong to them as individuals but the costs, in terms of grass depletion through grazing and manure pollution, fall on the community at large. While the overall cattle numbers are below or close to the capacity of the common no problems arise. Once this capacity is exceeded, however, milk yields will start to fall and the calves which are born will be leaner.

The tragedy lies in the logic which leads individuals to acquire more cows. This is relentless, even when problems have become severe. Six starving cows may produce less milk in total than two healthy ones, but seven will produce a little more than six, even if each is slightly hungrier. Hardin concludes that 'Freedom in a commons brings ruin to all' (Hardin 1968: 1245).

The power of this analogy lies in its ability to analyse the causes of environmental problems arising from small-scale interests abusing environmental capital to the detriment of larger-scale interests. Individuals who find it easier to drop litter than find a bin impose costs (disamenity and the cost of street cleaning) on their local communities. A community may prefer to pour its sewage into a river, producing problems for itself and other towns and villages, rather than pay for waste water treatment. Nations can and do use the atmosphere as a sink for pollution, producing trans-boundary and global problems (Boxes 1.2 and 7.1). In each of these cases the benefit accrues to the smaller-scale interest whilst the costs, either in direct financial terms or in depletion of environmental capital, are shared between larger-scale interests. This is the essence of the tragedy of the commons model.

The model can also be used to consider ways out of the tragedy as Box 2.2 seeks to illustrate. Hardin himself suggests various ways around the problem. One is privatisation: to create ownership also creates an incentive to manage the resource and prevent over-use. Although this is theoretically possible for some resources (such as National Parks threatened by too many visitors) for others it is more difficult (for example the atmosphere). Hardin dismisses the idea of exhorting individuals to behave responsibly and then relying on their consciences to make them behave as unrealistic. Instead, he advocates 'mutual coercion, mutually agreed upon': that is, exhortation backed up by sanctions, including legal sanctions.

Hardin's main theme is global population and he uses the model to argue that 'freedom to breed is intolerable' if it is the wider society (either within nations which are welfare states, or between nations in aid-donor/aid-recipient relationships) who will have to feed the children of the improvident. He thus denies the right of families to choose how many children they have – and this is a right enshrined in the Universal Declaration of Human Rights.

Box 2.2

The Framework Convention on Climate Change

Growing scientific evidence of anthropogenic climate change during the 1990s (Box 1.1) led to the establishment in 1990 of intergovernmental negotiations under the auspices of the United Nations (UN). By 1992 a Framework Convention on Climate Change (FCCC) had been drafted, ready to be launched at the Earth Summit of world leaders in Rio de Janeiro (see Chapter 7). The FCCC, as its name suggests, sets out a framework for future international negotiations rather than binding targets for emission reduction. Its objective is 'to achieve stabilisation of greenhouse gas concentrations in the atmosphere at a level that would prevent dangerous anthropogenic interference with the climate system within a timeframe sufficient to allow ecosystems to adapt naturally to climate change, to ensure that food production is not threatened and to enable economic development to proceed in a sustainable manner' (FCCC Secretariat 1992).

The convention is based on the twin principles of equity and common but differentiated responsibilities. From the outset it has been recognised that the developed nations resemble the householders in Hardin's model with many more cows out on the common, in terms of their CO_2 emissions per capita, than their less developed 'neighbours'. If developing nations were to cut, or even to stabilise, emissions it would severely curtail their ability to generate wealth and deal with the problems of poverty and increasing population. Stated that simply, the principles are clear. Translating them into the complex arrangements needed to establish a binding emissions regulatory regime at the global level has involved several years of negotiation and demonstrated somewhat unsteady progress.

Annual Conference of the Parties (COP) meetings have been held since 1995, although much detailed work takes place between meetings. The 'parties' are those countries that have ratified the convention. COP 1, held in Berlin in 1995, started negotiations on binding targets which were finalised at COP 3 (1997) with the Kyoto Protocol.

The Kyoto Protocol sets emission targets for developed nations (the so-called Annex 1 countries) in respect of the six main greenhouse gases – carbon dioxide, methane, nitrous oxide, hydrofluorocarbons, perfluorocarbons and sulphur hexafluoride. Using a base year of 1990 for CO_2, CH_4 and N_2O (1995 for the other three gases) and a compliance window of 2008–2012, the decreases (or constrained increases) for the Annex 1 nations are set out in the table below.

The wide range of targets reflects the differing circumstances of the parties to the convention. Australia, Iceland and Norway pleaded successfully to be allowed to increase emissions over the period as they were already using a large proportion of renewable energy and would find it difficult to continue to grow their economies without increasing fossil fuel use. Russia and the Ukraine had seen much of their energy-intensive industry collapse shortly after the fall of communism and are likely to show substantial emissions reduction over the Kyoto period without taking any specific action at all. But these states negotiated for a stabilisation target only, for reasons that will shortly become apparent.

continued

Box 2.2 continued

Emission targets relative to 1990 levels (per cent) proposed in the Kyoto Protocol for Annex 1 countries

Party	Target (2008–12)	Party	Target (2008–12)
Australia	+8	Liechtenstein	−8
Austria	−8	Lithuania	−8
Belgium	−8	Luxembourg	−8
Bulgaria	−8	Monaco	−8
Canada	−6	Netherlands	−8
Croatia	−5	New Zealand	0
Czech Republic	−8	Norway	+1
Denmark	−8	Poland	−6
Estonia	−8	Portugal	−8
EU	−8	Romania	−8
Finland	−8	Russian Federation	0
France	−8	Slovakia	−8
Germany	−8	Slovenia	−8
Greece	−8	Spain	−8
Hungary	−6	Sweden	−8
Iceland	+10	Switzerland	−8
Ireland	−8	Ukraine	0
Italy	−8	UK	−8
Japan	−6	USA	−7
Latvia	−8		

Source: FCCC Secretariat (1997)

When calculating their performance against these targets Annex 1 countries will be able to count land use projects which result in net absorption of carbon dioxide against their actual emissions, as long as these are linked with afforestation, reforestation, and deforestation since 1990.

As well as emissions reduction within their own frontiers, Annex 1 nations could co-operate with other countries in two ways under the Kyoto Protocol:

- Emissions trading (see Chapter 8) allows countries that exceed their targets to buy the unused emissions of countries who undershoot theirs – hence Russia and the Ukraine will find any shortfall between their actual and target emissions most lucrative. This has been called the 'hot air' mechanism. By removing incentives from both Russia, and any Annex 1 country preferring to pay rather than take the steps necessary to actually cut emissions, it weakens the protocol overall.
- Joint implementation is designed to reward Annex 1 countries that enter into partnership each other, or with developing countries, to fund and facilitate technology transfer or land use projects that result in net emissions reduction. The funding country will be able to set the calculated emissions reduction in the other country against its own actual emissions. For the offset to be recognised, however, it must be demonstrated that the funding, the activity and the emissions reduction are

Box 2.2 continued

all additional to what would have happened anyway, either through normal market forces or through existing aid or environmental programmes. This is called the additionality criterion. In practice, additionality may prove difficult to apply, thus further weakening the ability of the protocol to produce real, rather than notional, emission reductions.

Having given teeth to the FCCC at Kyoto, fleshing out the final details proved a tortuous process. Negotiations at COP 4 (Buenos Aires, 1998) and COP 5 (Bonn, 1999) were due to conclude at COP 6, held in The Hague at the end of 2000. The US delegation was extremely wary of making commitments because the presidential election was imminent. It proved impossible to find a package of measures that was acceptable to all parties. The meeting broke up amid rancorous exchanges between the European and US parties, with Britain caught in the middle.

In March 2001 the protocol was dealt a potentially fatal blow when the newly inaugurated US President George W. Bush announced that the United States would not ratify the protocol. During the 1990s, the 4.7 per cent of the world's population who lived in the United States were producing nearly a quarter of global CO_2 emissions, and these were predicted to grow rapidly into the twenty-first century. This unilateral action provided a clear disincentive to other participants to bear the cost of cutting their own emissions whilst the United States was free-riding – gaining economic advantage from unconstrained use of fossil fuels whilst potentially benefiting from any global environmental advantage of future reduced warming.

In fact, the reconvened COP 6 in July 2001 did manage to reach further agreements on the implementation of the protocol, although Australia joined the United States in stating it would not ratify the protocol. Operational details were finalised at COP 7 and COP 8, held in Marrakesh (2001) and New Delhi (2002). To come into force, however, the protocol must be ratified by at least fifty-five parties to the convention, including countries responsible for at least 55 per cent of Annex 1 carbon dioxide emissions in 1990. At the time of writing 104 countries had ratified, but, in the absence of the United States and Australia it was essential for the Russian Federation to join if the emissions condition was to be met. The Russian parliament was due to consider this early in 2003.

References: FCCC Secretariat (1997); Tata Research Institute (2002).

Discussion points

1 To what extent has the atmosphere been transformed from an open access regime to a commons regime since 1990?

2 Is the Kyoto Protocol an example of 'mutual coercion, mutually agreed upon'?

His line of argument is developed in a starker form by a further analogy, the lifeboat model (Hardin 1974), in which the relative status of the nations of the world is compared to the aftermath of a shipping disaster. Some are safe and comfortable, in lifeboats with sufficient resources to ensure survival. Others are swimming in the sea in grave peril. The conscience of those in lifeboats might lead them to take the swimmers on board, but, argues Hardin, that would be a mistake. Overloading the lifeboats might mean that all eventually drowned, through capsize, or died from lack of food and water. The conclusion drawn is that the developed nations can best ensure their own survival by denying aid to the developing world, especially those parts of the developing world where action to curtail population growth is ineffective or absent.

Despite their power and utility Hardin's models have been much criticised. One reason is Hardin's inaccurate use of the word 'commons'.

> what [Hardin] is describing is not a commons regime, in which authority over the use of forests, water and land rests with a community, but rather an open access regime, in which authority rests nowhere; in which there is no property at all; in which production for an external market takes social precedence over subsistence; in which production is not limited by considerations of long-term local abundance; in which 'people do not seem to talk to one another'; and in which profit for harvesters is the only operating social value.
>
> (*Ecologist* 1993: 13)

There are many historical and present-day examples of commonly held resources being managed effectively and efficiently. In medieval England grazing was organised much as described by the model, and the escalating over-use predicted was not an inevitable consequence. Today, many examples exist where common resources are exploited in a controlled way, despite the opportunity for private benefit to occur at the expense of public costs (*Ecologist* 1993; Gibson *et al.* 2000). The key conditions for successful management of common resources are summarised in Table 2.1.

However, by defining Hardin's model as an open access regime because of the absence of community control over exploitation of the common resource, these criticisms, in fact, serve to reinforce, rather than deny, the strength of Hardin's argument. Once the inaccurate nomenclature has been conceded, it can be seen that successful commons regimes work precisely because there is 'mutual coercion, mutually agreed upon'. Where this does not exist, in open access regimes, over-use and tragedy ensue just as the model predicts.

The real inadequacy associated with the tragedy of the commons and lifeboat models lies not in their use as a predictor of economic behaviour in the absence of social control, but in the failure to make explicit the moral assumptions which underlie their basic premises. As the *Ecologist* quotation above implies, the tragedy of the commons model presumes a

Table 2.1 *Attributes of successful common property regimes*

1   User groups need the right to organise, or at least no interference when they do
2   The boundaries of the resource must be clear
3   The criteria for membership of the eligible group of users must be clear
4   Users must have the right to modify their use rules over time
5   Use rules must correspond to what the eco-system can tolerate and should be conservative, to provide a margin for error
6   Use rules need to be clear and easily enforceable
7   Infractions of use rules must be monitored and punished
8   Distribution of decision-making rights and use rights to co-owners of the commons need not be egalitarian but must be viewed as 'fair'
9   Inexpensive and rapid methods are needed for resolving minor conflicts
10 Institutions for managing very large systems need to be layered, with considerable devolution of authority to small components to give them flexibility and some control over their fate

Source: McKean (2000) in Gibson *et al.* (2000): 43–9

capitalist economic system based on private property. Inequality in access to the basic resources necessary for survival is the starting point of the lifeboat model, but there is also an implicit assumption that the developed nations have only enough to guarantee their own survival. As the discussion of postmodern consumer culture has made clear, this is far from the case. Sitters are not assumed to have the ability to grow thinner in order to make space for some of the swimmers. Swimmers are assumed to be helpless, with no opportunity to save themselves without help from the sitters.

A truer analogy would have to include a few survivors in luxury yachts and many others in well provisioned cruise liners, as well as the lifeboat dwellers and swimmers. The mutual coercion necessary to regulate open access regimes must have as an aim the conservation of environmental capital – however, if it is truly to be mutually agreed upon, fairness and equity must also be its guiding principles. Hardin's analysis misses this basic point.

Attitudes and values

The moral dimensions of Hardin's models, therefore, must be taken into consideration when applying them to real life situations. One further important aspect is the values held by the commoners. What sort of people would act in such a way as to ruin the common resources on which they and others depended? Some adjectives which spring to mind are: selfish, materialistic, short-sighted, ignorant and narrow-minded.

Commoners who were altruistic, community-minded, long-term thinkers, wise and open to the ideas of others would almost certainly manage to avoid the 'inevitable' tragedy which Hardin predicts. Environmental policy makers therefore need to understand the relationship between values and the behaviour of others – and of themselves.

Values as a source of conflict

In Chapter 1 resources were defined in terms of their value and wastes in terms of their lack of value. Used in this sense the word 'value' means the worth of some object or attribute. This can sometimes be measured in terms of money, although money itself is valued more highly by some individuals than by others. But value is not an objective or absolute quality. There can be great variation between different individuals, different cultural and religious groups, different communities and different nations in the values they attribute to the same thing. Box 2.3 exemplifies the type of dispute in which there can be no consensus unless the value systems of the participants begin to converge. Interestingly, this case study suggests that disputes can sometimes precipitate such changes, albeit small and over a long time period.

Box 2.3

Coronation Hill

Coronation Hill, in the remote South Alligator region of the Northern Territory of Australia, contains very rich deposits of gold, palladium and platinum. There has been mining activity in the area since the 1950s and in the 1980s exploration began in preparation for large scale extraction of mineral resources. This aroused great controversy because of the opposition which arose from the Aboriginal Jawoyn people who were successful in their campaign to prevent further mining in the area.

The Jawoyn beliefs are not well documented as they are considered secret by the tribe. However, they consider many sites in the region to be sacred 'Bula sites'. The high value they place on these sites derives from the story of a god who, having created the earth and the human race, was injured by a hornet and became very sick. Crawling very many kilometres from site to site, he eventually entered the ground, becoming incorporated into the rocks. Ngan-mol is a term used to describe both the god's blood and the mineral ores within the rocks. If the god is disturbed he will wake and split the Earth in his rising: if not he will remain there for ever.

Jawoyn predictions of catastrophe formed the basis of their objections to further mining activity at Coronation Hill. Their claims were complicated by many factors. There was

Box 2.3 continued

uncertainty over the actual location of the Bula sites. The spatial characteristics of these were extremely complex, involving underground connections between sites and spheres of sacred influence reducing over distance, rather than precise boundaries. The unwillingness of the Jawoyn to break the taboo on discussion of the Bula myth may have disadvantaged them in establishing their claims. These factors were compounded by evidence that Coronation Hill had in fact been the territory of another people (the Wulwulam) and this led to suspicions that the Jawoyn had invented the Bula mythology fairly recently in order to expand their territory to include Coronation Hill. The fact that active religious observance of the Bula myth had ceased, although the sites remained taboo for many purposes, also added to the contentiousness of Jawoyn claims.

Although they complicated the resolution of this dispute, none of these factors represents the basic cause of the conflict between the Jawoyn and the mining operators. This arose because of the differing values that the two groups brought to the question of whether or not to mine at Coronation Hill. The mining companies were, of course, motivated by the economic value of the resources they hoped to extract from Coronation Hill. Other Australians, including some Aboriginals, supported the mining proposals because of the economic wealth that would be generated thereby.

For the older Jawoyn, however, the hill was infinitely more valuable undisturbed. The site had intrinsic value, because of its sacredness, but the prime reason for this allocation of value was the fear of the catastrophe that they believed would occur if the god and his Ngan-mol were disturbed. The question then arises of whether monetary compensation would meet the concerns of the Jawoyn to an extent great enough for them to withdraw their opposition. Other mining disputes in the Australian outback have been resolved in this way, with some Aboriginal communities receiving several millions of Australian dollars in royalties or compensation. In the case of Coronation Hill no such deal was struck and the mining operation was stalled by Prime Ministerial edict.

Jacobs (1993) suggests, however, that it is imprecise to characterise the values of the pro- and anti-mining lobbies as completely polarised. One of the effects that this and other mining disputes have had is to expose both sides of the conflict to the conceptual framework used by the other side. There is now much greater awareness of Aboriginal values and attitudes throughout the majority Australian population, for example. The difficulties of comprehension and interpretation that academics, administrators and politicians have encountered whilst attempting to resolve the issue show that full understanding is far from being achieved, however.

References: Jacobs (1993); Young (1995), chapter 5.

Discussion point

1   This is an extreme example, but diverse values underlie many environmental conflicts. Browse newspapers and news Web sites to find three further examples where the value placed on environmental assets by the protagonists is widely different. Is there evidence that the conflict is leading to better mutual understanding – or to entrenched positions and hardened attitudes?

Extrinsic and intrinsic values

So far in this chapter, the reference point for considering the importance of the environment has been human-centred. Only values assigned to environmental objects or attributes by human individuals within the human economy have been discussed.

It can be argued, however, that there are two types of value – extrinsic and intrinsic. *Extrinsic value* arises from the relationship between the valued object and the valuing observer. It is, of course, essentially subjective, arising from the value system of the observer. However, some analysts (e.g. Naess 1988) maintain that the environment and its component parts have *intrinsic value*. This means objective value which exists whether or not humans decide to consider the object valuable – or whether or not humans exist. Thus the component parts of the environment (individual organisms, entire eco-systems or even inanimate objects such as rocks) have value irrespective of whether they meet human needs, directly or vicariously.

Intrinsic value is associated with the set of beliefs known as *ecocentrism*. Ecocentric thinkers believe that:

- the human race is part of nature: its status is as one species amongst millions in the natural world;
- the natural world has an intrinsic value and that some parts of it are inviolate – for example the remaining wildernesses;
- the resource and waste demands of human societies must be managed so that they remain within the limited capacity of environmental systems;
- the inappropriate use of science and technology may create more problems in the long term than are solved in the short term;
- that the welfare of the human race depends at least as much on environmental quality as it does on material well-being;
- that preservation, not conservation, is the appropriate principle to apply to the management of the environment. The aim should be zero, not just minimum, environmental impact.

Ecocentrism is often presented as reactive. The forerunners of what is today called ecocentrism were frequently to be found opposing the economic and political arrangements of the day as well as expounding ideas about humankind's place within nature. For example, in the thirteenth century St Francis of Assisi rejected the idea that God put animals and plants on the Earth solely for the purpose of human exploitation. He was also troubled by the then growing wealth and ostentation of the monastic orders. Against some opposition from within the Church, he founded the Franciscan order of monks as an alternative for those who wished to follow the religious life in genuine poverty, as required by the monks' vow.

In the twentieth century, ecocentric individuals and movements have been associated with a reaction against modern and postmodern patterns of production and consumption, for example the campaigns against industrial whaling, nuclear power and globalisation. O'Riordan (1995) has sub-classified ecocentrism into the 'deep environmentalist' and 'soft technologist' strands. Deep environmentalists will strike extreme positions in order to prevent environmental damage, whereas soft technologists emphasise low impact and community-based means of meeting human needs whilst respecting the inherent value of the natural world.

Ecocentrism underlies much Green philosophy and thus is often relevant in a practical way to environmental policy making, especially when differences in the values of different groups have resulted in conflict. However, it should be noted that environmental capital is usually defined in terms of extrinsic value as determined by human beings and their needs. When environmental problems arise through the over-exploitation or mismanagement of other species it is important to recognise that these are defined from an anthropocentric (human-centred) viewpoint and not, usually, from consideration of the intrinsic value of other species or environmental attributes. The environment is presumed to be of value for human purposes alone. This is because policy making is, after all, a human activity and the anthropocentric allocation of values underlies this. This point of view is defined as *technocentrism*.

Technocentrism is a world view which:

- is based on anthropocentrism, as it sees humankind as separate from, and superior to, the natural world;
- believes that humankind has the right to manage the natural world in the cause of progress – i.e. the improvement of the material well-being of the human race;
- has faith in the ability of humankind to use science and technology, not only to progress, but also to resolve any environmental problems which arise as a result of progress;
- will tackle environmental problems from a conservationist angle – trying to extract the most environmental services for the least environmental damage.

As with ecocentrism, O'Riordan (1995) has identified two sub-strands within technocentrism. 'Accommodators' believe that environmental problems will need to be addressed seriously through effective environmental management and regulation. With this proviso, they foresee a future without significant limits to growth. 'Cornucopians' are even more optimistic, believing that environmental problems can always be circumvented through human ingenuity and the use of technology.

Values and the Greens

The mid-1960s saw the beginning of a rapid development of environmental awareness and concern around the world. Over the following decades, the environment became a mass concern, the subject of continuing interest from the public and political decision makers at every level, even though the level of such interest fluctuated somewhat. A clutch of new environmental groups emerged: some using traditional campaigning methods such as lobbying politicians, others willing to use more radical tactics such as demonstrations and direct action. By the late 1980s a significant number of individuals had begun to manifest concern for the environment, not only by joining some of the burgeoning environmental groups, but also by making consumer and lifestyle choices.

In the northern hemisphere these developments were strongly associated with changes in values that occurred during this time. Phases of rapid economic development during the twentieth century seem to have spurred increases in environmental concern, just as the industrial revolution did in the nineteenth (McCormick 1995). This concern can be measured by such indicators as the proportion of newspaper column centimetres given over to environmental issues or the number of environmental groups being formed during any one period (Lowe and Goyder 1983).

Inglehart (1977) made use of Maslow's hierarchy to compare the value systems of two different generations of Americans: those born before and those born after the Second World War. The impetus for the study was the clearly observable generation gap which emerged in the mid-1960s as the post-war generation came of age. Amongst this age group 'flower power', resistance to the Vietnam War and a new concern for the environment were all manifestations of values and attitudes very different from those of their parents. Inglehart developed a questionnaire which could be used to compare the attitudes, and by inference the values, of interviewees. The questions were designed to distinguish between materialist values (those concerned with the lower levels in Maslow's hierarchy) and post-materialist values (those concerned with upper levels). So, for example, questions concerned with economic growth, inflation and national defence were used to assess the value assigned to materialist factors by interviewees. Adherence to post-materialist values was revealed by questions concerned with democracy, relationships, community and quality of life.

The results of Inglehart's survey showed that, in both age groups, materialist values predominated, but that post-materialist values were much more widely held amongst those born after the war. In an apparent paradox, he also found that post-materialist values were also held more often by those who were more affluent, those who were in higher socio-economic groups and by those who were more highly educated. In fact these results can be accounted for by assuming that people will value most highly the things that they lack and undervalue the things that they have. The material security flowing from the unprecedented wealth and abundance of consumer goods in post-war America allowed people to downgrade the priority they gave to material values and to develop post-materialist values, in line with Maslow's model. In the UK Cotgrove and Duff (1981) adapted Inglehart's research methods and found that members of environmental groups were more likely than the general public to show adherence to post-materialist values. This raises the interesting and paradoxical possibility that concern for the environment may be a product of affluence.

Obviously the shift towards post-material values was important in catalysing the development of the modern environmental movement, but growing evidence of environmental problems associated with the increased throughput of resources in the human economy was an associated factor. Which of these played the prime role is a matter for debate: they are anyway likely to have interacted. Some analysts (for example Grove-White 1993) claim a third factor is involved. This is the uneasiness within modern consumer societies about the powerlessness of the individual in the face of technological and social change, which has latched on to the physical manifestations of environmental concern to find expression.

Within the modern environmental movement itself there is, of course, a range of philosophies, values, beliefs and attitudes. One characterisation that is often used distinguishes between deep and shallow Green approaches (Naess 1973; Pepper 1996: 17–34):

- *Deep ecology* is a fundamental approach based on an extreme ecocentric philosophy. Deep green thinking gives the highest priority to the health of the planet, with human considerations being secondary.

- *Shallow ecology*, whilst recognising the importance of the environment and the need to act to solve environmental problems, is more anthropocentric in its approach. The principle underpinning shallow green philosophy is the need to save the planet as a habitat for the human race, not the need to save the planet *per se*.

Other values interact with those found on the spectrum between extreme ecocentrism and extreme technocentrism and thus produce different types of environmentalism. The left/right divide in mainstream politics exists also in the environmental movement and is reflected in differing views held on, for example, the balance between individual liberty and collective responsibilities and the appropriate role of the government within the economy. Extreme ecocentrism, by down-playing the needs of humanity and therefore issues of equity in access to environmental services, has been labelled right-wing by some, more anthropocentric, analysts (e.g. Bookchin 1990). Environmentalism has intersected in a similar way with the politics of gender in the development of eco-feminism. This form of environmentalism blames patriarchal values and structures for the degradation of environmental capital and the oppression of women (e.g. Daly 1987). It is interesting that the connections between racial issues and environmental problems have been much less well analysed, although Shiva (1994) has expanded the eco-feminist critique to include consideration of colonialism and globalisation.

Values and policy making

Some of the discussion of values and value systems above may seem abstract and remote from the reality of environmental problem solving. However, the values that people hold will have consequences for the choices that they make and the behaviour which results. When policy solutions are sought for environmental problems, values are thus of considerable practical importance both in tragedy of the commons type problems and more complex examples of environmental conflict.

Policy makers need insight into their own value systems and those of other individuals and groups if they are to be able to develop policies which will be effective in changing human behaviour. Indeed, some environmental policies are designed to bring about changes in individual or corporate values as a means of changing behaviour to reduce the pressure on environmental capital, for example the CAMPFIRE project described in Box 3.1. However, there are countervailing forces at work. Advertising which seeks to engender desire for a particular product amongst consumers is an obvious example of the manipulation of values in a way that that will encourage consumption and waste production.

Valuing the future

As was noted earlier in this chapter, one fundamental difference between *Homo sapiens* and other species is the ability to envisage the future. However, this capacity can achieve future benefits only if it is combined with a belief that action now can make a difference and a willingness to take such action. In any set of cultural circumstances these factors will be problematic. The balance between immediate and future gratification can be difficult for individuals to achieve in their own lifetime or between themselves and their children. This is especially the case for those in poverty or close to poverty who may logically prefer to eat their seed corn, and starve next year, rather than starve at once. When people are asked to bear costs now in order to benefit 'future generations' who are not yet born and to whom they are not necessarily related, the disincentive to do so should be obvious. Where the risks to future generations are large, as is the case with nuclear waste, a balance can be found between present and future generations as Box 8.1 will demonstrate. In that example, the risks from nuclear waste produced over the last fifty years can be minimised, at considerable expense to electricity consumers, but not entirely eliminated.

However, for late twentieth-century Western cultures there are additional reasons to behave badly towards the future. These stem from pessimistic and fatalistic beliefs (revealed in visions of the future in popular culture) which are likely to engender the narrow and short-term attitudes evident in Hardin's commoners. Science fiction books and films do present some optimistic visions – for example, of nature throughout the universe conquered through technology for the benefit of humankind (*Star Trek, Star Wars*). If life is not sweet, it is for political, not environmental, reasons and a happy ending is a realistic prospect. More frequent (especially if the target audience is adult, not family) are visions of society and environment ravaged and diminished by the continuation of the postmodern late twentieth-century trends in production and consumption (*Blade Runner, Brazil*). Escape back to nature is an unrealised dream (at least in the *Blade Runner* Director's Cut, if not in the original version). Other films go further by predicting outright apocalypse through technology (*Mad Max, Terminator, Waterworld*). These latter two classes of films exemplify the loss of faith in progress and fear of the future which are among the symptoms of postmodernity.

Visioning the future is done in less popularist ways also. Governments and influential think-tanks make it their business to extrapolate trends in order to predict and plan for the future. The IPCC reports on global warming (Box 1.2) are just one example of this widespread activity. Yet often predictions of future environmental problems do not produce policies which might avert these. One reason for this in democratic societies is that future generations have no votes. Governments subject to four or five-year election cycles are reluctant to prejudice their electoral support by implementing policies which impose costs now in return for future benefits, even if those benefits may be manifest within the decade. Disadvantaging the here-and-now generation for the sake of people not yet born is an even less appealing task. An added complication is that prediction always entails some degree of uncertainty (see Chapter 4) and this can be used to prevent or delay action where evidence about environmental problems is not clear-cut.

Digging away at the roots

The causes of environmentally damaging behaviour are wide-ranging: some are inherent in human nature, others entrenched in cultural circumstances. Values and value systems are key, but, although significant changes in values have occurred and will continue to occur as modernist values give way to postmodernism, this is not a managed process. If, as Cotgrove's work suggests, significant material affluence is a prerequisite of environmentalism, the global prospects of timely action to reverse the depletion of environmental capital identified in Chapter 1 are poor.

The general goal of environmental policy has already been identified in terms of getting people to change their behaviour. But how can appropriate and inappropriate uses of environmental capital be differentiated? What is needed is an overarching set of principles to guide policy making. In Chapter 3 the use of sustainable development and other concepts to provide viable aims for environmental policy is introduced and evaluated.

Further reading

Ponting (1991) is a salutary read and tells the story of humankind's relationship with the environment since its evolution as a species. The best references on socio-biology are those cited in the text. Lyon (1999) is an accessible undergraduate text on the cultural shifts described in this chapter; Klein (1999) describes their impact and detects a nascent backlash. Hardin's 1968 original article on the tragedy of the commons is widely included in anthologies of environmental classics. In Gibson et al. (2000) several commons regimes across the world are examined and the features of sustainable management identified.

The philosophy of environmentalism and the values associated with the Green movement have been widely written about, for example Pepper (1996), Radcliffe (2002), Sutton (2000) and Benson (2000). Some aspects of religious influence on environmental attitudes are covered in Chapple and Tucker (2000). The ethics and practicalities of inter-generational equity are well covered in Dobson (1999).

3 ▶ Sustainable development and the goals of environmental policy

- 'Limits to growth' and 'sustainable development' as appropriate goals for environmental policy responses
- The meaning and implications of sustainable development
- Approaches to assessing sustainable development policies
- Alternative and less rigorous standards such as Best Practicable Environmental Option

Introduction

So far the main focus of this book has been consideration of the nature and causes of environmental problems. Starting here, chapters will now introduce the role of environmental policy in formulating and implementing possible solutions. The first thing that any policy maker needs to decide upon is goals. Chapter 3 therefore discusses how appropriate aims and objectives of environmental policy might be developed and what principles might underlie these.

In the first part of this chapter, various conceptions of what has been described as 'the predicament of mankind' are discussed, showing how ideas such as 'limits to growth' and 'sustainable development' can be used to set the aims of environmental policies. However, actual policies based on these principles are rarely put into practice in the real world, although in the case of sustainable development there is growing evidence of attempts to do so.

More often, policies will be less radical and aims will be constrained by the limits of what is deemed to be achievable, or practicable, in any given set of circumstances. Towards the end of the chapter, the use of these more limited approaches to environmental goal setting is discussed.

The predicament of humankind

Malthusianism

Although, as discussed in Chapter 2, technocentric attitudes have predominated throughout the development of industrialised societies, some minority thinkers have taken other approaches to the examination of the relationship between environment and humankind. One such was Thomas Robert Malthus (1766–1834), who was one of the first economists. In 1797 Malthus published the first version of the work for which he is now most famous, *An Essay on the Principle of Population*. Although subsequently subject to revision in response to the considerable critical response that greeted it (Malthus 1803), the central argument of the work remained the same through all editions. This was that human populations tended to grow exponentially whereas the food supply (agricultural output) could at most grow only arithmetically.

Thus Malthus argued that if a population doubled in twenty-five years (say from 1 million to 2 million) food supply might also double (from enough to feed 1 million to enough to feed 2 million) as population pressures forced improved productivity upon farmers. However, given a situation where the land most suitable for agriculture was already under cultivation, these productivity rises would be limited to what could be achieved by increasing labour inputs. The extra (marginal) returns from this increased labour would diminish year by year as land of lower fertility was brought into production and the extra work (weeding etc.) in cultivating the crops produced yields close to the maximum achievable.

In the second twenty-five-year period, if unchecked, the population would double again (to 4 million) and then again in each subsequent twenty-five-year period (to 8 million, then 16 million and so on). Yet the most optimistic assumption that Malthus was prepared to make about agricultural production was that it might be increased by the same fixed amount in each time period (to enough food for 3 million at the end of the second twenty-five-year period, then 4 million, 5 million etc.). He was of the opinion that even increases at this rate were unlikely, except in underpopulated regions, such as the New World was at the end of the eighteenth century. Even these areas, however, would inevitably become overpopulated, with respect to the food supply, in time.

Thus scarcity was the fate of mankind and the prime cause of it was not inefficient and unfair social and political arrangements, as some writers, for example Thomas Paine and others associated with the French and American revolutionary movements, were suggesting. According to

Malthus, the inevitability of poverty for the poorest sections of society was due instead to the combination of the laws that he had identified and which were both natural and divine. Human nature (as ordained by God the Creator) would lead to geometrical population rise and this would be checked only by the limits imposed by the nature of the resources God had put at humankind's disposal – that is, by starvation, infant mortality and other fatal diseases caused by want. In later editions of his essay Malthus softened this message somewhat by acknowledging that 'moral restraint' might also act as a check on population (Winch 1992). By moral restraint he meant (poor) individuals lowering the birth rate by remaining chaste and voluntarily postponing marriage, thus reducing the time span between marriage and menopause.

Malthusian arguments are, of course, highly ecocentric, making use of 'natural' rules to explain the existence of poverty and misery, and to a large extent dismissing the (technocentric) possibility of humankind managing nature, or even managing itself, to escape from its predicament. For example, Malthus advocated reform of the English Poor Laws, which enabled automatic financial support for families in dire poverty, because he believed that this encouraged population growth amongst the destitute.

Arguments against the Malthusian view can be divided into two broad categories. Cornucopians (Chapter 2) would claim that his analysis was incorrect with respect to food production. This argument is bolstered by examination of the 200 years since Malthus wrote his essay. These have seen global population rise exponentially (Figure 1.5). Although poverty and want have affected significant proportions of the human population during this period the food supply has on the whole risen, not arithmetically, as Malthus suggested it would, but exponentially in line with overall demand. Whether such rates of increase can continue into the future is discussed in Box 4.2.

Other opponents of Malthus argue that he was wrong about the inability of humankind to overcome the problems of population and distribution of resources by social and political means. Such arguments (e.g. Chase 1980) often highlight Malthus's background as a relatively affluent clergyman and academic who accepted the existing distribution of wealth and indeed often championed the interests of the land-owning classes. For example, Malthus favoured the retention of the Corn Laws, which discouraged the importation of grain and kept domestic prices (and therefore landowners' profits) high, even though in other matters he was an advocate of free trade (Winch 1992). It was bias such as this, it is claimed, that led Malthus to concentrate on the amount of food produced, rather than question the fairness of distribution arrangements. Empirical evidence does seem to show that poverty increases the incentive to have large families whereas increasing material well-being often leads to increased birth control (Leisinger *et al.* 2002). This is the precise opposite of Malthus's predictions.

The Limits to Growth

Malthusian arguments can be extended to consider the limits of not only food production but also other resource provision and waste systems. This approach underlay the influential report *The Limits to Growth* (Meadows *et al.* 1972). The authors used a computer model to examine the implications of the exponential growth of world population, and the associated growth in agricultural production, use of natural resources, industrial production and pollution.

The results of the most widely reported scenario, the reference case, are shown in Figure 3.1, which shows a rapid collapse in global economic output in the early years of the twenty-first century. To quote the legend of the figure in the original text:

> The 'standard' world model run assumes no major change in the physical, economic, or social relationships that have historically governed the development of the world system. All variables plotted here follow historical values from 1900 to 1970. Food, industrial output, and population grow exponentially until the rapidly diminishing resource base forces a slow down in economic growth. Because of natural delays in the system, both population and pollution continue to increase for some time after the peak of industrialisation. Population growth is finally halted by a rise in the death rate due to decreased food and medical services.
>
> (Meadows *et al.* 1972: 124)

Because the first restraining factor on economic growth in the reference case was the availability of natural resources, popular interpretations of *The Limits to Growth* at the time of publication tended to focus on resource depletion. This aspect of the study attracted much criticism: time has shown that the authors made insufficient allowance for the discovery of new reserves. For the reasons given in Chapter 1 the resource shortages predicted in *The Limits to Growth* have not arisen. In fairness to the study, several variations of the model presented were based on a more optimistic view of future resource availability. These, for example, Figure 3.2, tended to show rising pollution in overloaded waste sinks, causing an increase in the death rate and damaging food production, to be the limiting factor.

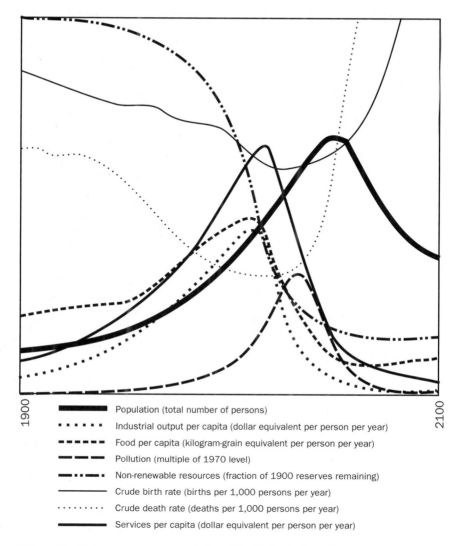

Population (total number of persons)

• • • • • • Industrial output per capita (dollar equivalent per person per year)

▬ ▬ ▬ ▬ ▬ ▬ Food per capita (kilogram-grain equivalent per person per year)

▬ ▬ ▬ ▬ Pollution (multiple of 1970 level)

▬ •• ▬ • Non-renewable resources (fraction of 1900 reserves remaining)

─────── Crude birth rate (births per 1,000 persons per year)

• • • • • • • • • Crude death rate (deaths per 1,000 persons per year)

▬▬▬▬▬ Services per capita (dollar equivalent per person per year)

Figure 3.1 *Reference case scenario from* The Limits to Growth

Source: Meadows *et al.* (1972: fig. 35), by courtesy of the Club of Rome, Hamburg

Even under what the authors claimed to be the most optimistic assumptions about resource availability, development of pollution control technology, availability of birth control and increases in food yield per hectare, industrial output and population grew exponentially until limited by one factor (or by a combination of these), peaked (in all cases before 2100), and then declined rapidly.

Other criticisms of the study (e.g. Kahn *et al.* 1976; Simon 1997) were more fundamental, for example:

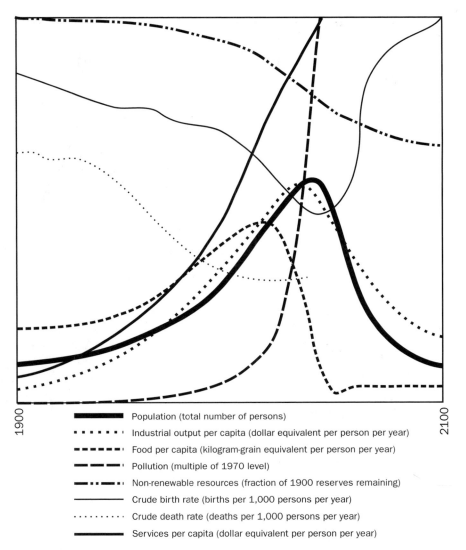

Figure 3.2 *Scenario from* The Limits to Growth *assuming the availability of unlimited resources*

Source: Meadows *et al.* (1972: fig. 37), by courtesy of the Club of Rome, Hamburg

- Its assumptions were too pessimistic about future technological advances which might ameliorate the problems it described, in particular the potential for increased efficiencies in energy and resource use.
- The computer model took into account too few variables and, by aggregating these to a global scale, grossly over-simplified the real world situation.

- The model did not distinguish between different types of economic growth. Growth can be energy and resource intensive and therefore damaging to the environment (e.g. growth due to an increase in road transport) or it can be environmentally benign (e.g. growth due to the expansion of well managed forestry industries).

However, the point of *The Limits to Growth* study was not to predict with any accuracy events in the twenty-first century: rather it was to demonstrate the 'exponential nature of human growth within a closed system' (Meadows *et al.* 1972: 189). The authors acknowledged that more resources and better technology could delay the time when the limits are reached but claimed that population growth linked to economic growth would, at some stage, be curtailed by the ability of the environment to sustain this. Even so, the authors clearly stated that the scenarios of economic and population collapse described above were not inevitable. It was their opinion (re-stated in the 1992 follow-up book *Beyond the Limits*) that:

> It is possible to stabilise these growth trends and to establish a condition of ecological and economic stability that is sustainable far into the future. The state of global equilibrium could be designed so that the basic material needs of each person on earth are satisfied and each person has an equal opportunity to realise his or her human potential.
>
> (Meadows *et al.* 1972: 24, 1992: xiii)

This 'state of global equilibrium' was based on stabilised population and material wealth (defined as capital invested). The only way to deal with the problems of global poverty and hunger in such a steady state world would be through redistribution of wealth. In order for the poor to become less poor, at least some of the rich, including those on average incomes in the developed world, would have to become less rich. Meadows *et al.* probably underestimate the political and economic difficulty of achieving equitable distribution of wealth in a steady state economy, but justifiably make the point that these difficulties would be much more severe in the catastrophic circumstances predicted by versions of their models based on continuing growth.

Limits to Growth as a policy goal

The idea that economic growth must cease as environmental limits are reached is not popular. Policies with aims derived from such a philosophy would be difficult to frame and to enforce. Even those who have access to sufficient resources to meet their basic needs still aspire to greater

material well-being, for themselves a
not have such access are condemned
policies, unless there is substantial r
course would be resisted by the rich
the redistribution of wealth, the mes
so much cake, and those who have r
slice must make do with the crumbs

Thus, acceptance of limits to growt
of environmental policy, but this ac
a world that contains deep inequalities. Policies based on this analysis (..
they could be made to work) would aim to achieve a steady state – zero
economic growth, with birth and death rates in balance. But the effects
of such policies on the material wealth of both rich and poor would be
severe. This would be true whether distributional policies were based
on Meadow's prescription of altruistic surrender of material wealth by
the rich to the poor, or on the very different approach advocated by
Malthus and Hardin (Chapter 2).

Therefore, whilst environmental limits are fundamental to environmental
policy making, they are, by themselves, an insufficient basis for policies
which are workable and fair. Consideration of distributional issues –
property rights, the fairness of access to resources and provision for the
poor – leads to policy prescriptions and policy aims which are more
sophisticated than those offered by the limits to growth analysis.

Sustainable development

The concept of sustainable development emerged from this conundrum.
The most commonly quoted definition of sustainable development is
taken from the 1987 report of the World Commission on Environment
and Development, otherwise known as the Brundtland report after the
commission's chair, Gro Harlem Brundtland. The following extract is
quoted at length to demonstrate the all-encompassing range of the idea,
which goes far beyond mere environmental policy.

> Humanity has the ability to make development sustainable – to ensure
> that it meets the needs of the present without compromising the ability
> of future generations to meet their own needs. The concept of
> sustainable development does imply limits – not absolute limits but
> limitations imposed by the present state of technology and social
> organisation on environmental resources and by the ability of the
> biosphere to absorb the effects of human activities. But technology
> and human organisation can both be managed and improved to make

for a new era of economic growth. The Commission believes that wide-spread poverty is no longer inevitable. Poverty is not only an evil in itself, but sustainable development requires meeting the basic needs of all and extending to all the opportunity to fulfil their aspirations for a better life. A world in which poverty is endemic will always be prone to ecological and other catastrophes.

Meeting essential needs requires not only a new era of economic growth for nations in which the majority are poor, but an assurance that those poor get their fair share of the resources required to sustain that growth. Such equity would be aided by political systems that secure effective citizen participation in decision making and by greater democracy in international decision making.

Sustainable global development requires that those who are more affluent adopt life-styles within the planet's ecological means – in their use of energy, for example. Further, rapidly growing populations can increase the pressure on resources and slow any rise in living standards; thus sustainable development can only be pursued if population size and growth are in harmony with the changing productive potential of the ecosystem.

Yet in the end, sustainable development is not a fixed state of harmony, but rather a process of change in which the exploitation of resources, the direction of investments, the orientation of technological development, and institutional change are made consistent with future as well as present needs. We do not pretend that the process is easy or straightforward. Painful choices have to made. Thus, in the final analysis, sustainable development must rest on political will.

(World Commission on Environment and Development 1987: 8–9)

What sustainable development means

Several issues fundamental to sustainable development emerge from a reading of the extract above. The first to note is that the widely quoted first sentence is a statement of faith rather than a statement of fact. Given humanity's record to date in meeting the needs of the present, let alone doing this in a way which shows respect for the needs of the future, it would be right to be sceptical of the realism of the aspirations expressed by the Commission. However, this is insufficient grounds on which to reject the concept unless there is a better one with which to replace it – and for the time being there is not. For by definition, there are only two alternatives to sustainable development – unsustainable development or no development at all.

Second, the concept of equity is fundamental to sustainable development. By equity, Brundtland means fairness in the distribution of environmental services within and between generations (so-called intra-generational and inter-generational equity) ensuring the basic needs of all are met. This is not the same as equality (everyone getting exactly the same) – but it is made clear that affluent lifestyles will have to adjust.

Third, the concept arises from a technocentric world view. People are seen as more important than nature. The needs of humanity are central to the definition and environmental limits are seen as negotiable through human ingenuity. The standstill in economic growth envisaged in *The Limits to Growth* analysis is not an acceptable option if the basic needs of the world's present and future populations are to be met.

Fourth, it is recognised that only through politics, and therefore policy, can sustainable development be achieved. Moreover, changes in the political systems that make decisions within and between nation-states will be needed.

Fifth, the definition is open to many different interpretations (Redclift 1987). What is meant by development – increased material wealth or a better quality of life? What are the legitimate needs of the present? What will be the legitimate needs of future generations and how can these be predicted? Does the responsibility for future generations literally mean forever? Most important, how will policy makers be able to measure whether their policies are truly sustainable or not? It could be argued that these imprecisions are one of the strengths of the concept – because they have allowed its acceptance by a wide range of inter-governmental, governmental and non-governmental organisations and because, by stimulating debate about the actual meaning of the concept, they have helped to publicise and popularise it. However, it should be clear that sustainable development as an aspiration needs further elaboration before it can be deemed useful to policy makers seeking to turn it into reality. The next sections examine some of the approaches that have been developed to help to apply the concept in a practical way. The case studies in Boxes 3.1–3 allow readers to apply these approaches in the contexts of wildlife and forest management in Zimbabwe, Finland and Madagascar.

Assessing sustainability: the three key criteria

One of the simplest ways to test policies and projects in terms of their compatibility with sustainable development is to apply the three key criteria of equity, futurity and valuing the environment (Pearce *et al.* 1989):

Box 3.1

Zimbabwe's CAMPFIRE project

However well the local environmental impacts of tourism are managed, the arrival of tourists in developing countries often brings social and economic pressures to bear on local communities. Environmental management practices which aim to preserve environmental capital may seriously interfere with the ability of native peoples to earn their living from the land. The creation of wildlife reserves where hunting is prohibited is a good example. Native people can find themselves excluded from areas in which previously they had open access to hunt and gather food and fuelwood. Furthermore, animals wandering beyond the reserve boundaries can predate on crops and cattle in nearby villages. Usually, the majority of the visitors are moderately wealthy people from developed countries. In some cases the desire of these travellers to enjoy the natural heritage of a different part of the world can result in a diminution of the standard of living of those whose needs were only barely being met in the first place.

Such conflict can be managed in various ways. One approach is based on the economic power of the two protagonists. The travellers have access to money; the local communities do not and therefore their interests count for little in the process. However, equity demands that such disputes should be settled by using an opposite logic, by putting the need for subsistence of poor people above the leisure pursuits of the affluent. Neither of these approaches actually resolves the problem in an acceptable manner. Putting the needs of tourists above those of local people can lead to illegal poaching and other environmental damage to nature reserves. If tourists are excluded from an area the developing country will lose the much needed foreign currency that they would otherwise have spent.

A more creative approach is to develop ways in which the situation can be managed to bring benefits to both groups whilst guaranteeing environmental protection at the same time. If wildlife can be managed in such a way so that the locals derive substantial material benefits from it, they are more likely to co-operate with tourism initiatives and the accompanying environmental management.

An example of this approach is the Communal Area Management Programme for Indigenous Resource Exploitation (CAMPFIRE) in Zimbabwe. Small-scale farmers in the arid areas bordering wildlife reserves regarded elephants and other large animals as pests. However, the potential financial gains from using land for safari operations were much greater than those from the existing agricultural practices. CAMPFIRE aims to support local communities with financial and technical expertise and by facilitating the introduction of legal and institutional structures which ensure that an appropriate proportion of the benefits of wildlife exploitation accrue to local communities.

One of the first projects established was set up by Nyaminyami Council in the Zambezi valley in 1986. A wildlife trust was created, funded by safari operations and game culls. This has proved profitable, with surpluses being diverted, via the council, into community development projects. Initial success now needs to be built on, by transferring skills brought in by outside NGOs in the early days of the project to local people to ensure the future independence of the project. It is also important to monitor

Box 3.1 continued

local attitudes to ensure that individuals who do not benefit directly from the operation continue to recognise the community benefits that derive from the trust's activities.

Reference: Toulmin *et al.* (1995).

Discussion points

1 From the limited information in the case study, do you think the CAMPFIRE project meets the three key sustainability criteria?

2 Which forms of environmental and social capital are mentioned in the case study?

3 Are there environmental impacts associated with tourism which are not mentioned?

● What will be the impact on the distribution of wealth, and other factors affecting quality of life such as environmental quality and exposure to crime or other hazards, between rich and poor? This is the *equity* criterion. Only policies and projects that are likely to reduce poverty and inequality are consistent with sustainable development.

● What will be the impact in the medium and long term? This is the *futurity* criterion. The principle of equity between generations means proposals with mainly short-term benefits and long-term costs, whether financial, social or environmental, should be rejected.

● Is the *value* of the environment taken properly into account? Proposals that treat environmental services as though they were free, and do not include consideration of how these costs will be minimised and mitigated, violate this criterion.

The criteria allow a broad set of judgements to be made, although answering the questions will often involve detailed assessments and evaluations. Inevitably, no matter how sophisticated the underlying analyses, judgements based on the three criteria will often be subjective and contentious.

Assessing sustainability: the capital approach

Environmental capital was introduced at the start of Chapter 1 as the source of environmental services. The term 'capital' derives from economics, where traditionally capital has been defined as the infrastructure (e.g. roads, power stations, factories) and equipment (e.g. machinery) which is used to produce material goods. When used in this

Box 3.2

Forest management in Finland

This case study assesses the sustainability of commercial forestry in Finland. By way of contrast, forestry in Madagascar is examined in Box 3.3.

In any forestry operation the sustainability of the timber resource will depend on the balance between felling and planting over several decades. But forests are much more than mere sources of timber: they represent many other forms of environmental capital, such as biodiversity and recreational resources.

The Finnish forest industry is of great national economic importance, contributing 26 per cent by value of the country's total exports in 2000. Seventy-one per cent of these forestry sourced exports are paper or paper products and 8 per cent pulp for paper and board manufacture. Finnish exports account for 15 per cent of world trade in paper and paperboard (FFIF 2001).

Finland is one of the most densely forested countries in the world with tree cover of more than two-thirds of the land area. This area is increasing, owing to tree planting on agricultural land. The total timber resource is 1927 million m^3: the growth of younger trees increases this by 75 million m^3 each year. For over 100 years Finnish forests have been managed using the principle of sustainable yield, ensuring that harvesting each year takes less than the replacement growth. Clear felled areas are reforested, sometimes naturally through allowing seeds present in the soil to germinate, sometimes by reseeding or planting saplings. Growing areas of trees are thinned as necessary to ensure maximum growth rate.

The graph shows that the rate at which timber was being regenerated is broadly in balance with the rate at which it has been extracted over this forty-year period, with a clear excess of regeneration over extraction emerging after 1975. This means that, applying the definition given in Chapter 1, the resource is clearly renewable and is being sustainably managed. This does not mean there is no adverse environmental impact from commercial forestry. The use of fossil fuels in planting, harvesting and processing needs to be taken into account, as does the use of pesticides (Box 1.1).

However, biodiversity is the most controversial aspect of this industry. For several years there has been a campaign (Finnish Nature League 2002) against the scale and the style of commercial forestry in Finland and other Scandinavian countries on the grounds of the adverse effects of timber extraction on wildlife and biodiversity, particularly in the ancient snow forests. Once land has been cleared for timber production, replanting results in a much less biodiverse mix of trees, plants and other animals. For example flying squirrels, whose habitat is the ancient mixed aspen forests found in Finland, are much less common than before because of habitat loss. The noise and disturbance caused by felling operations is also disruptive to some animal species.

Finland has introduced protection for approximately two-thirds of the ancient forests, with a plan based on landscape ecological principles. This means protected areas are linked by corridors of habitat suitable for indigenous wildlife to pass through so that

Box 3.2 continued

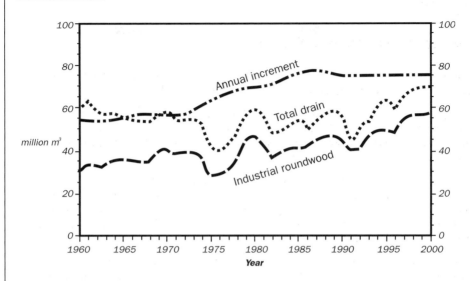

Forest balance in Finland, 1960–2000

Source: Finnish Forest Industries Federation (2001), by courtesy of the Federation, Helsinki

migration and interbreeding can occur. However, timber extraction is still permitted in unprotected areas, and, because the national forests have reached the limit of their productive capacity if felling is not to exceed new growth, Finnish firms have been establishing new operations across the border in Russia where the snow forests have been less exploited and the environmental protection regime is weaker.

References: Finnish Forest Industries Federation (2001); Finnish Nature League (2002).

Discussion points

1 From the limited information in the case study, do you think commercial forestry in Finland meets the three key sustainability criteria?

2 Which forms of environmental capital are mentioned in the case study? Is it possible to identify these as these as critical, constant or tradable? If not, what additional information would be needed to do this?

way capital only refers to things which have been human-made. Capital is one of the three factors of production recognised by traditional economics: the other two are labour (i.e. human work) and land. Land was traditionally defined as natural resources, but the term now more usually includes the much broader range of environmental capital reviewed in Chapter 1.

Box 3.3

Forest management in Madagascar

The majority of the population of Madagascar support themselves by subsistence agriculture. Forests are a resource in many senses. Amongst the most important uses are as land for growing food, as protection for lower ground from flooding, and as a supply of fuel wood, which is the main energy source on the island.

In the eastern rainforest area of the island the traditional method of agriculture is slash-and-burn. Areas of forest are cut down, then burned once the vegetation is dry. The land has limited fertility and so is cultivated for one or two seasons only, then left fallow. The farmer clears a new area, leaving the forest to regenerate the fallow area.

As the population is growing, more than doubling between 1960 and 1990, from 5.4 million to 12 million, there has been increasing pressure on the agricultural land resource and the length of time between cultivations of land patches can be as little as three to four years. Such a short rotation cycle cannot fully regenerate the soil's fertility, resulting in a spiral of decline: unproductive soils on the forest margin force farmers deeper into the forest to repeat the same cycle there. As cultivation moves from level landscapes on to sloping ground the added problem of soil erosion by rainwater run-off compounds the problem. This in turn is exacerbated by increased flooding due to deforestation.

As well as these agricultural pressures on the forests, the demand for wood as an energy source is also resulting in tree loss. Extensive deforestation has occurred: of the estimated original cover of 11.2 million ha, 7.6 million ha remained by 1950 and only 3.8 million ha by 1985. It has been predicted that all forest cover, save that on steep slopes unsuitable for cultivation, will be lost by 2025.

In Madagascar the fuelwood resource, soil fertility and flood protection potential are each declining. This decline could now be very difficult to reverse because of the loss of soil cover and soil fertility that has taken place.

References: Green and Sussman (1990); Kramer *et al.* (1997).

Discussion points

1 From the limited information in the case study, does forest management in Madagascar meet the three key sustainability criteria?

2 Which forms of environmental capital are mentioned in the case study? Is it possible to identify these as these as critical, constant or tradable? If not, what additional information would be needed to do so?

Chapter 8 introduces recent developments in environmental and ecological economics which have led to this redefinition. Also discussed there are the methods by which environmental capital and environmental services can be valued i.e. given a notional monetary value, and also the paradoxes and dilemmas which accompany these attempts. For the purposes of the discussion below it must be assumed that such valuation is a technical possibility: issues as to the validity of this assumption are examined in Chapter 8. The assumption is necessary at this stage because quantifying the value of environmental capital is one way in which policy makers can measure the sustainability of existing and proposed policies.

The concept of social capital was also introduced in Chapter 1, although the underpinning theory is much less well developed and measurement or economic valuation problematic (MacGillivray and Walker 2000). To the extent that social systems are meeting the needs of communities, whether these be for health care, child care, education, mobility, meaningful occupation or companionship, these systems can be considered as social capital. If they are damaged or allowed to deteriorate quality of life will suffer; if they are enhanced, quality of life will improve.

Sustainability as a principle requires that one generation should bequeath to its descendants at least as much capital as it inherited from its immediate ancestors. This obviously applies to all three forms of capital, economic, environmental and social. It is the responsibility of this generation to protect, maintain and enhance the economic infrastructure so that the next generation will have the same, or better, opportunities to produce material goods. Similarly, it would seem that sustainability implies the protection, maintenance and enhancement of environmental (and social) capital so that the flow of environmental (and social) services can be maintained or enhanced.

Environmental capital can be transformed into economic capital. The biodiversity of a wetland, for example, can be substituted for the economic capital of an industrial estate once the bog has been drained and built upon. Building the estate will also result in the depletion of non-renewable mineral resources: for building materials; making the machinery within the factories; and providing the energy needed for extraction of these resources, transporting them to the site and construction.

The substitutability of environmental capital for economic capital raises the important question of whether sustainability requires that it is the total amount of capital (economic, environmental and social) which should be maintained or enhanced, or whether each category should be considered separately. Is it acceptable to decrease environmental capital by a certain amount in order to enhance, to a greater extent, economic capital? The position that this is an acceptable way to apply the sustainability principle is called weak sustainability (Turner 1993). At the other extreme is the position of strong sustainability (see Figure 3.3). This maintains that, whilst it is fairly easy to convert a nature reserve into a motorway service station, it is far more difficult to reverse this action, and impossible in some circumstances, for example if species extinctions have occurred because of the loss of a rare type of habitat. It is easier to build up economic capital by polluting land, water and air than it is to restore these sinks to a pristine condition. Thus substitutability is mostly (but not invariably) a one-way process. Future generations are likely to be more populous than this one, at least throughout the twenty-first century. Their needs for environmental services cannot be predicted with any certainty. In order to maximise their chances to benefit from environmental services, policies for sustainability must therefore aim to preserve absolutely environmental capital.

However, neither of these extreme positions looks tenable upon a closer examination of the widely varying properties of different types of environmental capital. Insisting on the maintenance of every form of environmental capital in exactly the same condition as it is today would be basically a *Limits to Growth* position. Some forms of environmental capital are of greater importance than others and substitution of one form

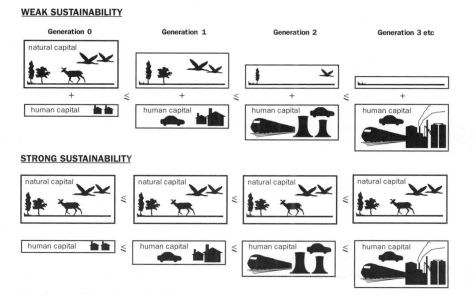

Figure 3.3 *Weak and strong sustainability*

of environmental capital for another is a viable option in some circumstances. One way of classifying environmental capital in a way that recognises these distinctions uses the categories of critical, constant and tradable (CAG 1997).

Critical environmental capital is the term used to describe attributes of the environment which are of the greatest importance. This may be because they perform vital functions (such as the stratospheric ozone layer which prevents most damaging ultra-violet solar radiation from reaching the earth's surface). Alternatively, the form of capital may be rare, but highly valued, such as the Bengal tiger and its habitat. For environmental capital to be defined as critical it must be irreplaceable. Questions of scale arise in the definition. A habitat that is rare in one region may be very common elsewhere. If it is valued sufficiently (for economic, ecological or aesthetic reasons) it may be designated critical at the regional scale, but not at the national. Once environmental assets have been defined as critical, the presumption for policy makers is that they must be at least preserved, and if possible enhanced, and handed down to future generations.

Constant environmental capital describes features that are also important, but where less stringent policies for their management are required. Constant assets are not yet scarce, but could become so if not managed carefully. There is presumed to be a threshold below which the stock of the asset must not fall, although the exact level of this threshold might not be known with any precision, or might be contentious. If the amount of this form of capital approaches threshold levels it becomes redefined as critical. The functions constant assets perform are important, but these could be fulfilled in other ways, either by replacement assets of the same type (extinction of a species in one area could be compensated by creating and enhancing habitats elsewhere) or by substitution with a different form of environmental capital (a grazing hillside may be destroyed by quarrying but a wildlife reserve or landfill site created after mineral extraction at the site has ceased). Spatial considerations apply to constant environmental capital in the same way as to critical.

The final category of environmental capital is described as *tradable*, as it is deemed acceptable to deplete this in order to create material wealth. Either such assets are not in danger of becoming scarce (such as oxygen in the atmosphere) or the environmental services they perform are not highly valued (for example rats and other vermin), or if they are, could be readily performed in other ways (loss of a recreational urban open space may be compensated by the creation of another near by; when suitable landfill sites become scarce, incineration of refuse is an alternative).

Categorising environmental capital as critical, constant or tradable is therefore a useful tool in theory for policy makers who need to decide which environmental assets must be conserved for future generations, and which are expendable. In practice, difficulties can arise when applying the concepts. An important set of problems concerns definition and these can occur whether it is the environmental function, its scarcity or the value of the asset that is being used to determine its category. These problems are introduced below and also raised by the discussion points of the Box 3.2 and Box 3.3 case studies.

Some forms of environmental capital perform a range of environmental services and may therefore be critical for some purposes and constant or tradable for others. An ancient oak tree standing alone on a village green may represent a constant environmental asset in terms of its biological value in its own right and as part of a set of eco-systems. In terms of the heritage and amenity of the village, however, it would be likely to be critical because the environmental service it provides could not be replaced for centuries. As noted above, designation at the margins of the categories may be difficult, owing to lack of scientific information about thresholds – this issue is examined further in Chapter 4. Even where good scientific information exists, its interpretation may be contentious. The debate between scientists and fishermen in almost every sea fishing area of the world over the level at which fish stocks should be maintained in order to guarantee their continuation is a good example of this (see Box 7.2).

Assessing sustainability: environmental space

The environmental space approach attempts to measure the sustainability of human lifestyles and so focuses upon human uses of environmental capital and environmental services. This emphasis on human activity and choice allows potentially a more sophisticated evaluation of the sustainability of processes and activities than the environmental capital model and one that is fully in line with the Brundtland approach, especially as it enables consideration of intra-generational, as well as inter-generational, equity.

Environmental space analyses start by calculating the sustainable rates at which environmental services can be consumed (McLaren et al. 1998). Scale is important in this calculation. For environmental services providing products that are traded internationally, for example minerals and agricultural goods, the global scale is appropriate. Some renewable resources, however, cannot be easily transported – water is a good example here. In these cases sustainable use must be calculated at the scale at which the resource is available.

Once the appropriate scale has been decided, the maximum rate at which the environmental service can be used is called the environmental space. The way this is calculated depends on the nature of the service:

- For *renewable resources* it is the rate at which the resource is regenerated.
- For *non-renewable resources* the rate depends on the abundance or scarcity of the resource, the environmental impact of its extraction, and the prospects for its replacement in due course by renewable alternatives.
- For *environmental sinks* it is the rate at which wastes can be assimilated without causing damage to other environmental systems.

Dividing the annual sustainable amount of the service by the population of the region, nation or globe (depending on the appropriate scale) gives each individual's 'fair share' of that particular resource – the amount which can be used without damaging environmental capital or infringing the rights of others to an equal share. The population figure used ought to reflect population growth predictions (see Figure 1.5).

Comparing actual consumption with environmental space targets highlights both over-use of environmental services and inequalities in access to these. Some calculations for the United Kingdom are given in Table 3.1. These are based on a 2050 target for achieving global sustainability and equity in the allocation of environmental space. This time scale was chosen on the basis that it allows sufficient time for developed countries to reduce their consumption and production patterns, and developing countries time to stabilise their populations (McLaren *et al.* 1998: 78).

Table 3.1 *Mid-1990s consumption figures and environmental space targets for some key resources*

Resource	Actual use (year)		Environmental space target, 2050	% cut needed
Carbon dioxide emissions (Mt)	576	(1990)	67	88
Aluminium (kt)	477	(1993)	59	88
Steel (Mt)	8.9	(1994)	1.53	83
Cement (Mt)	12.4	(1992)	3.45	73

Source: McLaren *et al.* 1998

An allied concept to environmental space is that of the ecological footprint (Wackernagel and Rees 1996). Whereas 'space' is the fair share consumption target for a particular resource, 'footprint' is the environmental impact of the actual consumption of all goods and services at a particular scale. The figures in Table 3.1 demonstrate the present outsize nature of the ecological footprint of the United Kingdom as a whole and demonstrate the scale of the changes needed to reduce this ecological footprint to what is sustainable.

Alternative goals for environmental policy

For these reasons, there is a plethora of approaches to environmental management offering aims and objectives that represent improvements in environmental performance but fall well short of the very demanding criterion of sustainability. The concept of Best Practicable Environmental Option (BPEO), which was written into UK law by the 1990 Environmental Protection Act, pre-dates the development of the sustainable development concept. The phrase BPEO was coined by the Royal Commission on Environmental Pollution in 1974 but is clearly based on the 'Best Practical Means' approach which was first used in Victorian air pollution legislation.

Best Practical Means and BPEO both involve balancing the need for environmental protection with technical and economic constraints:

> The aim is to find the optimum combination of available methods of disposal so as to limit damage to the environment to the greatest extent achievable for a reasonable and acceptable total combined cost to industry and the public purse.
>
> (RCEP 1985)

They share this characteristic with Best Available Techniques Not Entailing Excessive Costs (BATNEEC) which was also introduced into UK law by the 1990 Act. This is a regulatory approach, used for processes and substances capable of producing serious pollution, which requires polluters to install state-of-the-art equipment, and operate it in an environmentally optimal way, unless it can be shown that to do so would be unreasonably expensive for the company involved. Since the mid-1980s, however, there has been, in both the United Kingdom and the European Union (EU), a clear tendency to give increasing emphasis to the environment, and correspondingly less importance to the economic costs of action, when regulating industry using BPEO and BATNEEC. Indeed, BATNEEC has now been superseded by BAT (Best Available Techniques) under the European Union Integrated Pollution Prevention

Control (IPPC) Directive. This change removes the 'excessive' objection to installing the best possible operating plant in order to minimise environmental impact.

As well as taking the action forced upon them by a changing regulatory framework, businesses are also volunteering to improve their environmental management. Larger companies in particular have chosen to adopt environmental management systems such as ISO 14001, and are now pressing the smaller firms which supply them to do the same. All environmental management systems have as an aim continual environmental improvement. The theory and practice of an environmental management system is introduced more fully in Chapter 5 but for now it should be noted that the targets for improvement, and the time scale for meeting these, are set by the company itself.

Standards such as these have proved extremely useful in raising awareness within the business community about environmental issues and have resulted in very positive improvements in the environmental performance of some businesses and industries. It must be stressed, however, that compliance with legal requirements such as BPEO and voluntary standards such as ISO 14001 does not equate with sustainability, either in the short or the long term. This is because the criteria they apply are much less demanding than that of sustainability. At the time of writing a number of projects were under way to develop sustainability standards for organisations (e.g. BSI 2001) but none has yet been developed beyond the guidelines stage.

First, the focus of the standards is exclusively upon environmental impact and even that is often narrowly defined. For example, although the EU IPPC directive has resulted in many more companies in the United Kingdom than previously being subject to BAT, as well as tightening the criterion by replacing BATNEEC with BAT (National Society for Clean Air 2002), many firms are not subject to the standard. Some of these will be companies with huge indirect impacts on the environment and quality of life of people around the world, such as banks and other financial institutions. Others may be firms that have a large but diffuse environmental impact, such as airlines – IPPC applies only to products and processes with the potential to produce severe local damage. Companies not subject to IPPC may choose to adopt an environmental management system, of course.

Second, BPEO, BAT and ISO 14001 can all be applied without raising fundamental questions about the inherent sustainability (as opposed to the environmental impact) of a firm's processes and products. The human side of sustainable development – meeting the genuine needs of present

ons in a fair and efficient manner – is outside the remit
company manufacturing torture equipment for use by
leveloping world could claim full adherence to every
ital management standard but that would hardly
tainable.

evitably start from where the business is now and
...prove in the immediate future. This encourages incremental
adjustments towards better environmental performance rather than the
radical change necessary to move towards sustainability. As legislators
and regulators continue to strengthen their approaches, however,
environmental management standards may emerge without this drawback.
It could be argued that the much wider remit IPPC has given to UK
regulators (the Environment Agency (EA) in England and Wales; the
Scottish Environmental Protection Agency (SEPA) in Scotland), is
already evidence of this shift. Under IPPC regulated installations have to
conform to resource efficiency and waste minimisation, as well as
pollution and waste management, requirements.

From setting goals to scoring . . .

assessment from here,

Limits to Growth, as a goal for environmental policy, is inherently
unattractive as it fails to incorporate human aspirations for a continually
improving quality of life. Sustainable development, by contrast, offers a
fusion of environmental responsibility with a concern for the needs of all
people, those alive now and those yet to be born. This has resulted in a
challenging and powerful idea which will undoubtedly shape the
environmental policy agenda for decades to come. *the coming*

Sustainable development is a concept that has been subjected to a
torrent of academic and political analysis and criticism. It is difficult to
define and therefore can mean very different things to different people.
It is even more difficult to translate into realisable policy goals which
governments, businesses and other organisations can adopt. If such
policies can be developed they will be difficult to implement, as
sustainable development, however defined, is a very challenging aim.
And once implemented, it may be difficult to evaluate whether or not
policies have brought about outcomes which are truly sustainable because
of the lack of consensus about the definition, which brings us back to
square one. It is therefore unsurprising that, for most of the time,
practising environmental policy makers find themselves using BPEO
and other environmental management standards which are inherently less
demanding.

Whatever the desired outcome, policies have to undergo a formulation process: this will be the subject of Chapters 5–7, which consider in turn the corporate, national and international policy-making processes. However, the fundamental roles of scientific information and technological developments in environmental policy making must be first considered in Chapter 4.

Further reading

The World Commission on Environment and Development (1987) report sets out the Brundtland vision of sustainable development – it is a historically important document, accessible and worth reading. For a critical review of the origins and post-Brundtland development of the concept in the context of the developing world see Adams (2001). Selman (1996) explores sustainability issues at the local level. Environmental space and ecological footprint are well explained in McLaren *et al.* (1998). The UK Countryside Agency site <http://www.qualityoflifecapital.org.uk> demonstrates the practical applicability of the capital approach to sustainability. Hart (1997) discusses the limitations of environmental management based on standards such as BPEO and how and why companies can move towards sustainability.

④ Science and technology: policies and paradoxes

- The relationship between scientific information and policy making
- How policy can be made when scientific information is uncertain
- Technology in the context of environmental problems and sustainable development
- The concepts of appropriate technology, ecological modernisation and preventative environmental management

Introduction

The emphasis in this book on human behaviour as the key determinant of environmental problems and environmental solutions does not mean that environmental science is irrelevant to environmental policy makers. Previous chapters have introduced concepts such as environmental capital, environmental services and environmental limits in abstract terms. Only scientific information about the functioning of environmental systems can make these ideas concrete by, for example, determining how much a forest's 'sustainable yield' might be (Box 3.2). But policy makers need to understand the limitations of the scientific method, as well as the information it generates.

Putting science into practical use through technology has been seen both as the cause of environmental problems and as the source of potential solutions. Technology exists to meet human needs – what are the characteristics of appropriate technologies that can meet these needs sustainably?

Science for policy making

What do policy makers need to know?

> *Environmental problems can be categorised into three broad categories.*
>
> (Dovers *et al.* 2001; Table 4.1)

Table 4.1 *Attributes and scale descriptors for framing policy problems in sustainability*

Problem-framing attributes

1	*Spatial scale of cause or effect*			
	Local	National	Regional	Global
2	*Magnitude of possible impacts*			
2a	*Impacts on natural systems*			
	Minor	Moderate	Severe	Catastrophic
2b	*Impacts on human systems*			
	Minor	Moderate	Severe	Catastrophic
3	*Temporal scale of possible impacts*			
3a	*Timing*			
	Near-term (months, years)	Medium-term (years, decades)	Long-term (decades, centuries)	
3b	*Longevity of possible impacts*			
	Near-term (months, years)	Medium-term (years, decades)	Long-term (decades, centuries)	
4	*Reversibility*			
	Easily/quickly reversed	Difficult/expensive to reverse	Irreversible	
5	*Mensurability of factors and processes*			
	Well known	Risk	Uncertainty	Ignorance
6	*Degree of complexity and connectivity*			
	Discrete, linear	Complex, involving multiple feedbacks and linkages		

Response-framing attributes

7	*Nature of cause(s)*		
	Discrete, simple	Fundamental, systemic	
8	*Relevance to the polity*		
	Irrelevant/beyond jurisdiction	Primary responsibility	
9	*Tractability*		
9a	*Availability of means*		
	Fully sufficient	Available instruments/ arrangements/technologies	Totally insufficient
9b	*Acceptability of means*		
	Negligible opposition	Moral/social/political/ economic barriers	Insurmountable opposition
10	*Public concern*		
10a	*Level of public concern*		
	Low	Moderate	High
10b	*Basis of public concern*		
	Widely shared	Moderate variance in understanding	Disparate perceptions
11	*Existence of goals*		
	Clearly stated	Generally stated	Absent

Source: adapted from Dovers *et al.* (2001: 5)

Micro-problems have the majority of problem characteristics at the lower end (left hand side) of the table. It is likely that appropriate responses will similarly be driven by descriptors from the left. Examples would be the management of a particular habitat or of a minor pollution incident. Most often existing institutions and policies (or incremental developments of these) will be able to resolve the problems. Meso-problems have characteristics from the mid-range of the ranges. Examples are regional or national level issues, perhaps related to forestry or public transport, which can be resolved at the sub-national or national level. Macro-problems have characteristics found at the upper end (right-hand side) of the continuum, examples being stratospheric ozone depletion (Box 5.2) and global warming (Boxes 1.2–3, 2.2). These will be found on international policy making agendas – new research and new forms of policy response are likely to be necessary before the problems can be tackled.

The table shows how the information needs of policy makers are for both social science (points 8, 9b, 10 and 11; Chapters 5–8 of this book) and natural science (rest of the table; this chapter). Ideally, policy makers faced with an environmental problem would have access to the following scientific information:

- What is the nature, severity and spatial scale of the problem?
- If no action is taken, how will the problem develop over time?
- Is the problem itself likely to cause further problems (e.g. will BSE in cows spread to other species)?
- What factors are causing the problem?
- Has environmental capital been damaged; if so is this damage reversible?
- If environmental capital is affected, is this critical, constant or tradable?
- What options are there to reduce or eliminate the problem; how much will they cost and how long would they take to work? How effective will they be?

Box 4.1 describes, for the case bovine spongiform encephalopathy (BSE), one example of the complex science which can underlie environmental problems and the difficulties which ensue if full and complete answers to the above questions are not available to policy makers at the time they needed to make key decisions. This circumstance is not the fault of either science or scientists. The capacity of science to deliver information about environmental problems is inevitably less than ideal. This is in part due to methodological difficulties. But the relationship of science and scientists to society in general and policy systems in particular is also problematic.

The strengths and weaknesses of the scientific method

Reductionism

Reductionism is an approach to scientific methodology, which was developed in the early seventeenth century by the French philosopher and scientist Descartes. The reductionist approach involves explaining the operation of a system by the mechanistic functioning of its component parts. Application of reductionist methods of inquiry have led to great advances in scientific knowledge, but even closed systems cannot entirely be explained in terms of the isolated functions of their components. For open systems a simple reductionist approach is even less helpful, as an excessive focus on the system components may lead to external inputs into the system being relatively disregarded. Even within systems, synergy between different components will not be identified if each component is examined in isolation from the others. The whole is very often greater than the sum of its parts but this will not be recognised unless it is the whole, and not just the parts, which is investigated.

Controversy over the Gaia hypothesis, first proposed by James Lovelock, illustrates well the philosophical and practical problems that are generated by the mis-match of reductionist and holistic approaches to science. It is universally accepted by scientists that living organisms have the capacity to self-regulate their metabolisms using a process called homeostasis. For example, mammalian species have metabolic systems to keep their body temperature, and the volume and electrolytic content of the blood, within certain narrow ranges. If these systems fail to work the animal will die. Lovelock's novel proposal was that the Earth as a whole also has this self-regulatory ability. He named the concept and the planet Gaia, after the ancient Greek Earth goddess.

> In many ways Gaia . . . is difficult to describe. The nearest I can reach is to say that Gaia is an evolving system, a system made up from all living things and their surface environment, the oceans, atmosphere, and crustal rocks, the two parts tightly coupled and indivisible. It is an 'emergent domain' – a system that has emerged from the reciprocal evolution of organisms and their environment over the eons of life on Earth. In this system, the self-regulation of climate and chemical composition are entirely automatic. Self-regulation emerges as the system evolves. No foresight, planning, or teleology (suggestion of design or purpose in nature) are involved.
>
> (Lovelock 1991: 11)

Despite a long and eminent scientific career prior to the formulation of the Gaia hypothesis, Lovelock encountered exclusion by the scientific community when he tried to promulgate his theory. Publication in prestige journals such as Science and Nature was denied to papers about the hypothesis authored by Lovelock and his collaborator, the biologist Lynn Margulis (Lovelock 1991: 24). Lovelock sees this as a symptom of the stranglehold that reductionist ways of thinking have upon scientists, rendering them unable to contemplate holistic interpretations of nature. Not only does this mean that their work cannot be informed by holistic views; it can also lead to barriers between different scientific disciplines so that physicists, for example, understand little of biology, and oceanographers little of biochemistry.

Positivism and falsification

In the late nineteenth and early twentieth centuries a group of European philosophers known as the positivists developed the position that knowledge can be gained only through experience and empirical knowledge of natural phenomena. Religious and metaphysical belief systems are therefore invalid. An influential, leading positivist, Karl Popper (1965), characterised the scientific method as a continuing process of hypothesis proposal, testing, then either rejection or re-testing of the hypothesis. Popper maintained that a hypothesis can be valid only if it is testable, for it is only by testing that false hypotheses can be identified and refuted. Hypotheses that are tested and not refuted may stand: they are valid for the time being but not necessarily true, as future testing may yet falsify them. The logical conclusion is that, whilst it is possible to show hypotheses to be false, it is impossible to prove that they are true. This means that verification of any hypothesis is a logical impossibility.

For our poor policy maker, seeking from science answers to specific questions, this is very bad news indeed. For example, public safety was a crucial concern in the case study outlined in Box 4.1. The electorate wants to know if existing practices are safe or if they are dangerous. The politician wants similar information, reluctant to incur the costs of what might be unnecessary action to slaughter all herds where BSE infection has been detected. So the politician asks the civil servant to ask the scientists, 'Is it safe?' But the scientists can never truthfully answer, 'Yes, 100 per cent', to that question.

No current evidence of harm does not mean that evidence may not be out there but as yet undetected, or that evidence may not arise months, years or decades into the future, as happened with BSE. There may be a hypothesis that a chemical released to the environment will have a particular and damaging effect. There may even be statistical information

Box 4.1

Bovine spongiform encephalopathy

Spongiform encephalopathy causes deterioration in the tissues of the brain leading to neurological disorders and, eventually, death. It has been recognised in many mammalian species, including humans – the human form is Creutzfeld Jakob disease (CJD) – but until recently it was unknown in cattle. The first confirmed case of bovine spongiform encephalopathy (BSE) occurred in Kent, in 1985. The epidemic peaked in 1992 when over 35,000 cases were reported.

Once the disease was identified, policy makers in the UK government had two important questions for scientists: what was causing BSE; and was public health at risk from meat from infected cows? The first was resolved fairly quickly. Examination of histories of cattle with BSE showed that all infected cows had eaten prepared feed derived from animal meal, i.e. the remains of sheep, cattle and other animals. The evidence that this was the means of infection was strong enough by July 1988 for the government to introduce a ban on feed containing animal protein for cattle, sheep and deer – a straightforward example of science informing policy.

The danger to human health from consumption of beef was more difficult to establish. In the late 1980s and early 1990s the official line was that there was no evidence that BSE could cause CJD in humans and that therefore British beef was safe to eat. The clear priority of ministers was to protect the economically important beef industry from the catastrophic damage that would occur if BSE was believed by domestic and international consumers to be transmissible to humans. However, a time lapse of several years was to be expected between infection and the onset of symptoms. No evidence of harm could possibly emerge until the (unknown) incubation period of the then hypothetical form of 'human BSE' had elapsed.

In support of its position the government quoted scientific experts. Yet Sir Richard Southwood, who chaired a scientific advisory committee, has claimed that his recommendations were not based solely on his assessment of the current scientific knowledge but were tailored by his perceptions of what would be acceptable to senior civil servants and ministers. As Winter observes: 'here we have a situation where the politicians could claim, on scientific grounds, that a ban [on the use of cattle offal in human food] was not necessary whereas the scientists only came to this conclusion because of their perceptions of the political unacceptability of a ban' (1996: 563). In fact in November 1989 a ban on offal from cattle of more than six months of age being used in human foodstuffs and pharmaceutical products was introduced as a result of growing consumer concern about the safety of British beef.

In March 1996 it was announced that a new type of human CJD had been detected. The symptomology of this was significantly different from that usually seen in CJD cases. For example the mean age of CJD sufferers had been sixty-three years, but the ten victims of the new type had an average age of twenty-seven years. Scientific advice to ministers was that it was probable (but not proven) that these new cases (called variant CJD or vCJD) were due to beef consumption by the victims in the mid to late 1980s.

continued

Box 4.1 continued

Two new scientific questions now arose and science was unable to offer unequivocal answers to either. First, if BSE was responsible for these illnesses, had the unfortunate victims contracted the disease from offal (long banned from the human food chain by 1996) or from prime beef or even milk or cheese? The policy implications of these alternatives were obviously very different. Biologists had established that prions were not detectable in muscle tissue or milk, so it seemed unlikely that these foods were the source of infection, but available knowledge was not sufficient to rule out the possibility entirely.

Second, how many more cases of vCJD might emerge? Five years on there was still little clarity on this question. In November 2001, with 111 UK cases of vCJD diagnosed, a study by Valleron et al. claimed the most likely final toll would be just above 200, with a maximum limit of 405. In January 2002 worst-case predictions of between 50,000 and 100,000 deaths from exposure to infected cattle were published (Ferguson et al. 2002), with the warning that if BSE had spread to the UK sheep flock a further 50,000 human infections and deaths might occur.

References: Ferguson et al. (2002); Valleron et al. (2001); Winter (1996).

Discussion points

1   At which points in the history of BSE could the precautionary principle have been applied? To what extent was it actually used?

2   A diet which is over-rich in the saturated fats found in beef and beef products is known to be associated with cardio-vascular disease, which kills more than 250,000 (40 per cent of all deaths) in the United Kingdom every year. Why then did the relatively small number of deaths from vCJD in the 1990s cause such concern amongst policy makers and the public?

showing a correlation between the factor (for example, and from Box 1.1, pesticides in the environment) and the hypothetical effect (decline in bird populations). But this does not prove that pesticides are damaging bird populations. Establishing causation to a sufficiently high standard of proof means not only suggesting a plausible mechanism for the effect (the hypothesis), but also finding scientific evidence to support the hypothesis. Not least of the difficulties will be agreeing between the parties what is 'sufficient' proof in these circumstances. The victims of alleged pollution are likely to settle for a much lower standard of evidence than are those whose profits depend on business as usual.

Scientists understand the distinction between correlation and causation well and will couch their reports to policy makers in terms of probabilities. Thus the IPCC predictions of sea level rise in Box 1.2 show average outcomes from different scenarios and models, with upper and lower ranges. This does not mean that outcomes beyond this range are excluded but that scientists regard them as very improbable.

Box 4.1 appears to show that politicians may well themselves understand these distinctions but are fearful that the concepts are too complex for the public to take on board. Thus the scientists' 'no evidence of harm' is very often translated into the politicians' public pronouncement of 'safe'. Furthermore, the case study suggests that the sensitivities of scientists working in a politicised environment can lead them to tone down unpalatable conclusions. It is perhaps fortunate, therefore, that the public will not necessarily believe all that they are told by politicians and by scientists.

Science in society

The emergence of postmodernism (Chapter 2) is relevant in two ways to the relationship between science and society. First, and in practical terms, the cultural changes described as 'postmodern', such as changes in production and consumption patterns, are accompanied by observed changes in attitudes. The transition from modern to postmodern is associated with decreasing respect for institutions (church, firm and government) and decreasing faith in the ideas such as democracy and science, which emerged in the Enlightenment and forged the modern age and the industrial revolution (Powell 1998).

Post-modern societies are 'risk societies' (Beck 1992; Grove-White 1993), pervaded by anxiety about the consequences of industrialisation. Despite (or because) of apparently increasing short-term security in terms of material well-being and longevity, worries about issues such as pollution, vaccination, nuclear weapons and genetically modified food are increasingly articulated. The inability of science to deliver unanimous, convincing and genuine reassurance, and politicians' mis-translations of scientific judgements about safety, fuels this process.

The second interface between postmodernism and science is philosophical. Post-modernist philosophy (e.g. Lyotard 1984) has developed an intellectual position that is directly opposed to positivism. The new theorists view science as a socially constructed activity operating as part of the power structure within society. Science is funded and directed by governmental and corporate vested interests, including the interests of the scientists themselves ('we need more research . . .'). For this reason alone the truths science claims to reveal must be partial. However, the deeper question is raised, does nature has an objective reality at all? All science is ultimately interpreted through human perception. Postmodernist thinkers such as Lyotard claim that no single account of these perceptions has greater validity than any other. Science therefore has an equivalent status to astrology; Western medicine and crystal healing have equal authority.

Although policy makers need to understand the basis of postmodern accounts of science, to adopt these would hardly be of practical help in solving environmental problems. Despite the limitations of science there is no belief system that has demonstrated a superior track record in revealing information about the physical and natural world. The challenge is rather to break down, where possible, the divide between citizens and experts. At every scale – from householders fearful of microwave radiation from nearby telecommunication masts to tropical nations being asked to forgo their right to exploit their forest resources (see Chapter 7) – scientific expertise can too easily be used to bulldoze arguments and proposals through, ignoring the concerns of potential victims (Boehmer-Christiansen 2001; Irwin 1995). Involvement of all stakeholders and acknowledgement of the methodological, philosophical and political features of the scientific endeavour will, in most cases, lead to a more democratic outcome than bland and dismissive reassurances.

Making policy when science is uncertain

If policy makers do not have satisfactory answers to the questions they pose, how then can they then proceed? Inactivity while scientists carry out the needed research is not a viable option when the potential risks are large. Two broad classes of response are possible in such circumstances – no-regrets strategies or application of the precautionary principle.

'No regrets' strategies

One useful side effect of the emergence of a previously unrecognised environmental problem can be to force a rethink of long established industrial processes. It may be possible to make improvements to the process, which eliminate or reduce the potential pollutant. Often there are associated reductions in production costs and/or improvements in product quality as a result. This approach is called a no-regrets strategy. The changes have unambiguous economic benefits, which are entirely unrelated to any environmental gains. Uncertainty over the role of the pollutant in causing the problem will not affect the decision, except where the economic gains are marginal or themselves uncertain. Cost-effective energy efficiency improvements are often cited as examples of the no regrets approach.

The precautionary principle

The dilemma for policy makers begins at the point when all potential no regrets strategies have been exhausted and the problem is not yet resolved. If scientific evidence is absent or uncertain, what justification can there be for interfering with the status quo, especially if the costs of doing so will be large for one or more stakeholders?

Pollutant A is being released into the atmosphere. Species B is now in decline and a small number of scientists are claiming that A may be the culprit, although they have not yet proved this conclusively. Imagine that in this very simplified scenario the policy maker has only two options: do nothing or ban A. One obvious point of reference in these circumstances is the principle of sustainable development and the three key criteria which underpin it: equity, futurity and valuing the environment. The equity question will mean identifying winners and losers for both options in order to assess which is the fairest – but without proof this cannot be done. The futurity criterion means doing the same balancing exercise between present and future generations – again impossible in the absence of good scientific information. Valuing the environment will mean identifying what type of environmental capital B represents – critical, constant or tradable? This begins to be more helpful. If B is tradable (i.e. neither scarce nor particularly valuable), it will be more difficult to justify a ban than if the capital is critical (scarce, valuable and irreplaceable). But even if B is categorised as critical environmental capital the difficulty remains that there is no proof that A is causing its decline.

In the absence of hard science each of these is a question of judgement and what underlies each judgement is the concept of risk. At its most basic level risk is something that is routinely managed in every day life. Before stepping off the pavement to cross the road, most people will look first. In advance of exotic holidays, inoculations are arranged against tropical diseases. No one expects their house to burn down, but most will pay a small annual insurance premium to ensure they will be compensated if it does. In each of these examples we are employing the precautionary principle – accepting a small and immediate price to minimise the risk of future costs which could be of far greater magnitude.

Applying this approach to the vexed issue of A and B clarifies the issues without removing the need for judgement. For each of the three criteria the most useful question is not what will happen but what might happen? Best and worst-case scenarios based on business as usual can be developed, and their probability evaluated. Applying the precautionary principle is then a matter of deciding whether it is worth paying the 'insurance premium' (the cost of banning A) in order to avoid this risk.

Where there is uncertainty about environmental problems policy makers can take either an optimistic or a pessimistic view (Figure 4.1). In either case time will tell and the outcome will lie somewhere between these two extremes. If optimism has shaped the policy, business will have carried on as usual and, provided the optimists were proved correct by events, the state of the world will be excellent – an economy unhindered by bans and regulations, coupled with an environment where the predicted problems have never in fact been manifested. If, however, the pessimists had been right all along, the result of business as usual will have been environmental disaster. Applying the precautionary principle means listening to the pessimists and paying the price of adopting their policies. If the optimists were right the future would be economically disadvantaged by these unnecessary costs, but with a healthy environment the overall position would be good. But if the pessimists were right the very worst of the environmental disaster will have been avoided. Thus the application of the precautionary principle minimises risk.

Box 4.1 is an interesting case study of the non-precautionary approach to policy making, viewed with the crystal clarity of hindsight. The optimists were believed and the pessimists were right, although the full extent of the resulting 'disaster' (Figure 4.1) is not yet known. If, as soon as the disease had been recognised, draconian action had been taken to exclude infected carcasses from the animal and human food chains, many thousands of BSE cases, and maybe most of the vCJD cases, might have been avoided. Such a decision might have been argued for on precautionary grounds,

	Optimists' policies adopted	Pessimists' policies adopted
Outcome if optimists right	EXCELLENT	GOOD
Outcome if pessimists right	DISASTER	TOLERABLE

Figure 4.1 *Risk and precaution*

Source: adapted from Costanza (1989), quoted in Pearce *et al.* (1989)

but, paradoxically, could not ever have been justified using scientific evidence, for without the twin epidemics the evidence of bovine to human transmission would never have existed.

Technology: sheep in wolf's clothing?

Technology is the 'appliance of science' and has just as significant a role in environmental policy formulation and implementation. This is demonstrated by even the most basic consideration of the relationship between technology, the natural environment and sustainable development. Although some animals use primitive tools such as sticks and stones, the development of technology, and enjoyment of the benefits that have flowed from this, has set the human species apart from all others.

These benefits are huge. It is technology that has allowed the global population to rise to more than 6 billion. Agriculture and irrigation were the founding technologies of the first civilisations and are the only viable way of meeting the food needs of the world. Medical and public health technologies such as vaccination, antibiotics and sewerage systems have decreased infant mortality and extended the life expectancy of adults. As well as meeting such basic needs, technology enables a proportion of the world's population live comfortable and affluent lives, with the opportunity to travel and enjoy consumer goods.

However, the benefits of technology come with a price. Technology not only allows human consumption of resources at a rate many times faster than any other species – it also allows people to use their ingenuity to escape the immediate consequences of the over-exploitation of local environmental capital. An animal population which outgrew its natural resource base, or damaged critical environmental capital by creating wastes faster than these could be assimilated, would suffer a sharp population decline and maybe extinction as a result. Because humans have the ability, through technology, to transport over large distances either themselves, or the goods they need, communities can develop beyond the limitations of any particular local environment – but not, of course, beyond global limits.

Technology and sustainable development

Technology shapes human systems as profoundly as these systems are themselves shaped by the application of technology. There are many ways of defining technology. For environmental policy makers the following definition is most appropriate: *technology is any activity by which humankind manipulates the environment, or resources extracted from the environment, in order to meet its needs.* Although it is focused upon the relationship between technology and the environment this definition is very broad. It encompasses very simple technologies, such as the stone tools used by early humans in the Neolithic Age, as well as twenty-first century technologies such as communications satellites. As well as including the development and use of individual tools and machines, it encompasses much broader activities such as agriculture and transport. The final phrase 'in order to meets its needs' could be argued to be redundant, but is included for two reasons. The first of these is that it is true (even technologies that can be viewed as antisocial, such as weapons of mass destruction, are satisfiers of human needs at one of Maslow's levels). The second is that it makes clear the relationship between technology and sustainable development – both are targeted at the satisfaction of human needs through the exploitation of environmental capital.

Technology, therefore, is a key instrument in meeting human needs and sustainable development cannot be achieved without it. But, whereas sustainable development is a normative concept, expressing values such as inter- and intra-generational equity, technology as defined above is morally neutral. Once technology is deployed, however, moral and political issues will arise. Almost all of the case studies boxed into chapters of this book are concerned with environmental problems caused by the use of technology.

On the face of it, this is paradoxical: technology appears simultaneously to be both the means of meeting human needs and the cause of environmental problems. On deeper examination, of course, it becomes evident that this is less of a paradox than first appears. The issue is not 'does technology assist or impede sustainable development' but rather 'which types of technologies are likely to assist progress towards sustainable development, and which will impede this?' Analysing the environmental, social and economic consequences of the deployment of different technologies will answer this latter question. The case study in Box 4.2 demonstrates some of the complexities involved in this kind of assessment. The principles underlying such an assessment are outlined in the remainder of this chapter.

Box 4.2

The Green revolution

During the 1960s and 1970s new breeds of high-yielding varieties (HYVs) of wheat and rice became widely adopted in developing countries. For example, dwarf rice varieties were developed which were able to produce five times the yield of traditional strains. Since the 1990s the development of new varieties has accelerated rapidly, owing to advances in genetic engineering. However, contrary to expectations, hunger is still widespread in some developing countries. There are two possible explanations. Either the HYVs have failed to deliver the expected increases in yield, or they have, but a significant number of local people have not derived benefit from this.

HYVs – an undelivered promise?

There is no doubt that increased yields have followed the introduction of HYVs in some areas. In India, for example, the production of wheat and similar crops has more than tripled since the 1950s. However, the gains have been unevenly distributed, even within countries. Regions with a history of agricultural reform seem better able to make the necessary adaptations to cultivation practices, some of which are fairly complex. New varieties have also performed better in areas where the soil fertility was not degraded.

Some early HYVs performed well only in tightly defined moisture conditions, so it was necessary to have irrigation and drainage infrastructure in place to take advantage of them. More recently HYVs have been bred to withstand aridity and flooding so that they can be used successfully by those without access to irrigation systems.

However, fertiliser, pesticides, agricultural machinery, as well as the seeds in the first place, are necessary if HYVs are to grow and crop to their full potential. Farmers with large land holdings will be able to purchase these additional inputs. Those with smaller holdings often cannot. This is for financial reasons, but is also due to social factors such as lack of education and inability to negotiate with higher-status urban bureaucrats for loans and training. As yields rise, so do land prices, making the expansion of smallholdings unviable. Thus the introduction of HYVs can exacerbate the divisions between rich and poor and leave the poorest farmers more vulnerable than before.

Latterly, the introduction of genetically modified crops has raised anew all the above concerns, along with some which are novel to this technology. The so-called 'terminator' gene, which prevents viable seed being harvested from crops, will have the inevitable consequence of binding farmers who use it ever closer to the multinational corporations that developed it. Crops designed to be impervious to certain herbicides will have the same effect, as the herbicide manufacturer is often the same company that developed the seed.

The environmental effects of the new cultivation methods are also questionable. The artificial fertiliser required is very energy-intensive to produce. Its application can lead to nitrate pollution of drinking water. The pesticides used may harm the health of local

continued

Box 4.2 continued

people, those who consume the food, as well as local eco-systems (Box 1.1). There are also concerns that the mass cultivation of only a few types of HYV is narrowing the genetic diversity of crops and depleting natural resistance to pests. Cross-breeding between indigenous and genetically modified crops may have adverse effects on the long-term biodiversity of the gene pool.

Can the Green revolution continue into the twenty-first century?

The rapid rate of productivity gains seen in recent decades as a result of the Green revolution now seem to be slowing. Although genetic engineering techniques offer the prospect of faster development of new varieties, some analysts believe that a productivity plateau may soon be reached. Some developing regions have already reached the situation where extra applications of fertiliser and other inputs result in only marginal increases in yields. This is due to the law of diminishing returns, originally developed by Malthus. New approaches will be needed if agricultural production is to keep pace with population growth in the twenty-first century.

There is scope for the development of HYVs of subsistence crops such as sorghum, cassava and millet. Advances in genetic engineering offer the prospect of food plants which adapt more easily than the old HYVs to extreme environments and which are not dependent on large inputs of fertiliser and pesticides. If these advances come to pass the Green revolution will continue to deliver the productivity gains necessary to feed the burgeoning population of the world: if not, Malthus may yet be proved correct in his analysis.

References: Abraham (1991); Myers (1994: 56–8); Rifkin (1997).

Discussion points

1   In Europe, and to some extent in the United States, there is concern about the environmental and health effects of genetically modified food products. This concern is much less widely expressed in developing countries. Which ideas in Chapter 4 help to explain why this might be so?

2   Are HYVs examples of appropriate technologies? Are GMOs examples of appropriate technologies?

Appropriate technology

Appropriate technology is a concept was first developed by Schumacher (1973) in the context of the technological needs of the developing world. but is equally useful when considering the developed nations. The three key criteria (equity, valuing the environment, futurity) are a useful starting point in assessing the appropriateness of technologies. Any evaluation must be comprehensive, considering the indirect as well as the direct impacts. Both the products of the technology and the processes by which these are produced must be included.

For any particular product the equity criterion can be framed in terms of its social usefulness and contribution to quality of life. Technologies applied to meet basic needs, such as food, medicine and energy, might be judged more appropriate than technologies producing what Max-Neef (1991) has called pseudo-satisfiers and others would label luxury goods, such as designer clothes, sports cars and private jets. Often there will be winners and losers: products offering benefits to one group of stakeholders will disadvantage other groups. New roads mean that traffic flows more quickly: a boon to commuters, but a curse to those living near by and affected by noise, air pollution and the risk of injury to pedestrians.

For processes the equity issues are less straightforward and control of technology is key. Increased production of food in developing countries, for example, will not necessarily meet the needs of the indigenous poor, as Box 4.2 demonstrates. Agricultural technologies which are small scale, require little capital, have low inputs overall and produce subsistence crops are more likely to maintain or increase equity than technologies which are large scale, capital intensive, depend on high inputs of energy or agricultural chemicals and produce commodities for export.

The valuing the environment criterion can also be applied to both product and process. Appropriate technologies will themselves be efficient in terms of resource and energy use and their products likewise. Material and energy inputs will be sourced sustainably, from renewable or recycled origins where possible. Wastes will be minimised in the production process. All components of the product will be designed for minimum end-of-life impact, so that reuse and materials recycling will be straightforward alternatives to disposal.

Relevant impacts on the future again include resource and waste issues, but economic factors are also important here because the market system favours short-term profitability and may ignore long term costs. This means that, left to the market, technologies with high initial capital requirements but large long-term benefits will not attract investment funds (e.g. Box 4.3). Technologies that offer short-term benefits but impose large costs on future generations are inappropriate. Appropriate technologies will not deplete the resource base available to future generations, either by exhausting non-renewable resources or by over-exploitation of renewable resources. Wastes from technological processes should either be non-toxic and capable of disposal without significant long-term environmental harm, or degrade and be assimilated within the space of a human generation.

Technologies that meet one criterion may violate another. Such conflicts can occur between the three different categories or even within them. The electricity generation industry provides plentiful examples of this. Tidal barrages such as the proposed Severn barrage (Box 4.3), for example, tap a renewable energy source, reduce carbon dioxide emissions, and would be an economic asset to future generations. But the short- and long-term environmental costs are high. Nuclear power (Box 8.1) might appear to score highly in intergenerational terms, as carbon dioxide emissions from the use of this technology are very low, thus reducing future global warming. However, the long-term nature of the technology's decommissioning and waste management requirements count against it. The economic benefits are taken by one generation but a large proportion of the costs, both economic and environmental, is devolved down the centuries.

The context is also important to the assessment. Neither of these capital-intensive technologies would necessarily be socially appropriate in developing countries with a poorly developed electricity distribution infrastructure. Although they would meet the needs of more affluent, urban dwellers, who are connected to the grid and who own refrigerators and other devices fuelled by electricity, they would be irrelevant to the energy needs of the rural poor. Investment in what Schumacher (1973) called intermediate technologies, such as small-scale hydro-power and efficient wood burning stoves, might be far more likely to meet the needs of rural dwellers, improve their quality of life and reduce inequalities.

Ecological modernisation

Ecological modernisation (Gouldson and Murphy 1997) is the term used to describe the recent shift towards a more holistic and preventative approach to pollution prevention and waste minimisation which, it is argued, brings associated benefits to both the environment and the economy. Promoting change in this way is said (e.g. Andersen and Massa 2000) to lead to innovation and technological developments with micro- and macro-economic benefits, such as improved competitiveness and employment gains, beyond the efficiency and environmental gains initially hoped for.

Box 4.3

The Severn tidal barrage

The river Severn estuary has the second largest tidal range in the world. The proposed Severn barrage would have sluice gates open to allow the water up stream with the incoming tide. The gates would be closed at high tide to trap water upstream as the tide receded on the seaward side of the barrage. Once sufficient head of water had developed, the water upstream would be allowed to escape through the turbines, generating electricity.

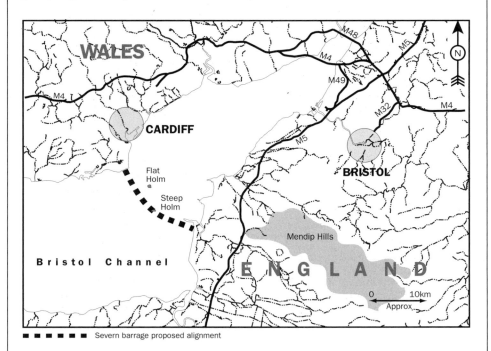

■ ■ ■ ■ ■ Severn barrage proposed alignment

The Severn estuary, showing the route of the proposed barrage

Source: adapted from Department of Energy (1989: fig. 1.1), by courtesy of HMSO, Norwich, Crown copyright

Energy output can be increased by using electricity to pump water through the barrage in an upstream direction at high tide, when the difference in water level across the barrage is small. A volume equivalent to this pumped water is released downstream later in the generation cycle, when the height difference across the barrage is large, therefore generating more electricity than was used to pump the water upstream.

A 1989 report showed that the project had the potential to generate 17 TWh of electricity annually, equivalent to 7 per cent of the then electricity consumption of England and Wales. However, the project is unlikely to be built in the foreseeable future for economic reasons.

continued

Box 4.3 continued

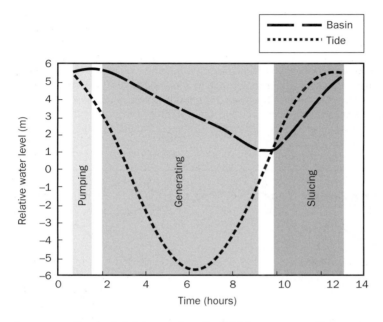

Pumping, generating and sluicing cycles of a tidal barrage operation

Source: adapted from Department of Energy (1989: fig. 2.10), by courtesy of HMSO, Norwich, Crown copyright

Local direct effects would be mainly due to the changed patterns of water movement in the estuary resulting from the construction and operation of the barrage. Physical systems within the estuary would be affected, both upstream and downstream: hydrodynamic systems (currents and waves); sedimentation systems; and salinity variation systems. This would result in changes in the pattern of intertidal areas (mudflats exposed at low tide and covered at high tide). The ecology of the estuary would be altered – the interdependent populations of plankton, invertebrates, fish and birds (waders and wildfowl) that the estuary supports would change in distribution and absolute numbers of species. The changes predicted are complex and not well understood and it is not possible to predict them with certainty.

However, the construction and operation of the barrage would also have environmental effects beyond the south-west of the United Kingdom. Some fish and bird species which inhabit the estuary are migratory: a decline, or increase, in their population, as a result of the ecological changes described above, would impact on population levels in their other habitats as well. For four bird species (curlew, dunlin, redshank and shelduck) the Severn estuary is an internationally important habitat and, as such, is protected under the Ramsar Convention. The eels which are found in the Severn have undertaken a three-year journey from their spawning beds in the Sargasso Sea between Bermuda and Puerto Rico, whence they will return to breed once they are mature.

Box 4.3 continued

The proposed structure is 15.9 km long and would consist of prefabricated caissons made from reinforced concrete, supplemented with embankments engineered at either end. Additional rock and sand would be required to engineer the foundations and to ballast the caissons. Some of this material could be obtained locally; the sand requirements of the project could be met by dredging in the estuary itself. However, significant proportions of the cement and aggregate would probably come from outside the Severn region. Some rock might have to be imported into the United Kingdom, possibly from coastal quarries in Africa.

One obvious environmental advantage of the barrage would be that, all other things being equal, it could replace the energy output of more polluting power stations using fossil fuel. The 1989 study predicted a resulting reduction in atmospheric carbon emissions of 4.8 MtC per annum, as well as reduced emissions of acid gases and other wastes. Changes in the fuels used for electricity generation in the United Kingdom since 1989 mean that these figures are now overestimates but significant emission reductions would be possible.

There would be many far-reaching effects on the economy of the region with knock-on environmental effects. For example, a proposed dual carriageway road on top of the barrage would result in development pressures along the link roads to the motorway network on either side of the barrage. The changed hydrodynamics upstream of the barrage would allow a large increase in the recreational use of the basin for sailing and other water sports. This would also lead to development pressure on the shoreline and hinterland of the estuary. It is likely the barrage itself would become a tourist attraction, generating traffic flows as well as electricity.

Reference: Department of Energy (1989).

Discussion points

1   Is the barrage project 'appropriate technology'?

2   How could environmentalist opponents of the Severn barrage scheme use demand management in their arguments?

There are three strands to the process. First, as industrial economies move away from the resource intensive industries that characterised the twentieth century towards service and knowledge based economic sectors, resource use and waste production per unit of economic output will fall. Second, incentives can be used to encourage the use of less polluting technologies, for example taxes on fossil fuels will make renewable technologies more attractive. Third, increased resource efficiency through waste minimisation and recycling will bring both economic and environmental gains.

The driver for this process is not consideration of appropriateness, but rather the enlightened economic self-interest of industry and government. For businesses, efficiency makes sense in economic, as well as in environmental, terms. Wasteful production processes are costly. They need extra resource inputs and excessive waste disposal costs. Getting more for less is a message any business can understand and approve of. Government can encourage ecological modernisation by switching taxes from income and employment to resources and wastes (see Chapter 8 for some examples). This will raise the cost base of inefficient businesses and put them at a disadvantage in the market place.

There would be obvious advantages if developing economies could be encouraged to avoid the mistakes made by the first world and skip, as far as is possible, the resource-hungry and inefficient stages of economic growth. However, infrastructure is crucial – only if there is access to facilities such as recycling systems, public transport and reliable networks for ICT connectivity can companies and consumers develop the new technologies of production and consumption that deliver more wealth for less environmental impact. Some analyses (e.g. McLaren *et al.* 1998: 256) claim that the shift to a service economy has not in fact diminished the developed world's environmental impact. Rather, manufacturing, and its associated unsustainable uses of environmental services, has moved to the developing world from whence manufactured goods are exported to meet the needs of first world consumers.

Preventative environmental management

Ecological modernisation requires resource efficiency, waste minimisation and pollution prevention to be designed into industrial processes from the outset, rather than considered only when pollution and wastes are causing environmental problems. Abatement, or end-of-pipe approaches, have been the conventional means of pollution control (Figure 4.2). When an actual or potential emissions problem is identified, control equipment is fitted to the end of the process to remove or treat the effluent before its release. Whilst this reduces the problem, there are inherent disadvantages to end-of pipe solutions (Jackson 1996):

- End-of-pipe technologies create knock-on effects, requiring resource inputs and producing wastes as a consequence of their operation.
- End-of-pipe technologies require energy to run and will lower the overall energy efficiency of the process.
- The installation and operation of end-of-pipe technologies represents an additional economic cost to industrial processes.

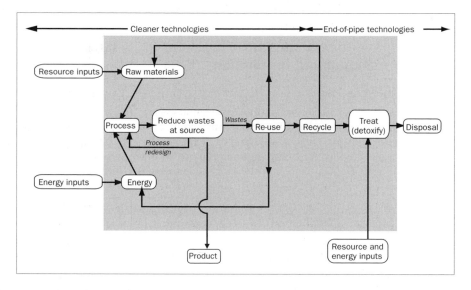

Figure 4.2 *Preventative and end-of-pipe approaches in a manufacturing process*

Box 4.4 illustrates the practical difficulties that can be encountered in waste management, which is an end-of-pipe activity. By increasing resource efficiency and reducing the toxicity of chemicals used through process redesign, or minimising waste through reuse and recycling of materials and energy, preventative approaches avoid these disadvantages. The key concept underlying preventative environmental management is that technologies should be as efficient as possible in their use of environmental services. The technique for assessing this efficiency is life-cycle assessment (LCA). This is a methodology that can be applied to either products or processes in order to quantify and compare environmental impacts. LCA is a two-stage process. Inventory analysis is the first stage. Starting with the extraction of raw materials that feed into the process or product, and ending with the final disposal of all wastes thus generated, all lifetime inputs and outputs are listed. The second stage is impact assessment. The environmental effects of each inventory item are categorised (in respect of, for example, global warming, ozone depletion, water quality) and quantified. The sum lifetime impact per unit of production can then be calculated for each category.

LCA analyses can be used when processes or products are being designed to minimise environmental impact. The concepts of Best Practicable Environmental Option (Chapter 3) and the waste management hierarchy (Figure 4.3) are used in conjunction with LCA. BPEO is, to a large extent, a common sense principle that can be used to identify from a set

Box 4.4

Waste management in Hamburg

Germany is a federal state, divided into twenty-one *Länder*, or states. Hamburg, as Germany's second largest city (1,662,000 population in 1991), is a *Land* in its own right. This has entailed difficulties for the city in formulating its waste management strategy. There is little land of rural character within the city boundaries, so most waste destined for landfill must be exported to other *Länder*, although there are waste incineration plants in the city and proposals to build at least one more.

During the period since the Second World War the volume and weight of Hamburg's waste stream has been rising, despite a population which has been in decline since the 1960s. This is because of rapid economic growth leading to increased consumption per head of population and therefore increased waste production, especially with regard to the quantity of packaging waste. The city's response in the 1960s and 1970s was to invest heavily in incineration capacity, some of which is now nearing the end of its useful life. In addition there is separate collection of putrescible waste for composting but these arrangements entail practical difficulties (contamination of organic waste by plastics and glass) and are very expensive to operate.

However, the future of the city's waste management is politically contentious for environmental, social and economic reasons. Neighbouring *Länder* are experiencing similar pressures of their own and are reluctant to allow Hamburg access to incinerators and landfill sites in their territories, these sites being limited in both size and number by environmental and planning pressures. As Hamburg is forced increasingly to deal with its waste within the *Land* boundaries, the costs to the city, which have to be passed on to its householders and businesses, are rising. Public concern about the environmental and health effects of incineration is growing, due in part to a campaign waged by the Green Party, which won 13.5 per cent of the vote in the 1993 senate elections. The Greens are also pressing for increased priority to be given to waste minimisation and recycling in the overall waste management strategy.

In 1989 the city administration proposed a solution to the growing crisis. This was to build a large waste to energy plant, taking 40 per cent of Hamburg's waste. The plant would generate both heat and power which would be sold by the private sector operators. These plans were strongly opposed, however, by the Green Party, other environmental and community groups, plus the trade unions, who were against the privatisation of waste management implicit in the incinerator plan. It was claimed that the diversion of such a large proportion of the city's waste stream to the new plant would reduce the incentive for waste minimisation and materials recovery. An alternative plan was put forward, based on waste minimisation and recycling, which would be backed up by increasing the capacity of some existing incinerators and seeking to develop new landfill sites outside the city boundaries. Although this would be a more costly option than that proposed by the municipality it is claimed to be more environmentally benign as it moves at least some of the waste stream to higher options in the waste management hierarchy.

Box 4.4 continued

Key points for consideration are:

* the difficulty of balancing economic, environmental and social considerations in formulating a strategy for waste management in Hamburg;
* the role of per capita consumption: even though the population is declining the waste produced in the city as a whole has been increasing.

Reference: Gandy (1994), chapter 5.

Discussion point

1 How could the municipality influence individuals and businesses in order to slow down or reverse the increasing quantities of waste?

of options those with the minimum combined economic and environmental costs. It is not a methodology, although attempts to apply the principle sometimes need to be based on complex methods, such as LCA.

The waste management hierarchy was developed as a decision-making framework (Figure 4.3 and Box 4.4). At the bottom of the hierarchy is the option of disposal – discarding the product and its component materials back to the environment. All other things being equal, this is likely to be the most costly option in terms of the use of environmental services. The resources from which the product was manufactured will be effectively lost from the economy and presumably be replaced by extracting primary resources from the environment. Environmental sinks will be burdened with the waste product for as long as it takes to be assimilated.

The next option is recovery, which means extracting economic value from the materials that comprise the waste product. Recycling or composting the materials in the waste to produce something of value or incinerating materials such as paper and plastic to produce useful energy can do this. On the face of it this approach reduces the environmental impact of the waste product. The need for primary resource extraction is minimised by the use of reprocessed materials or the displacement of fossil fuels to generate energy. Pressure on environmental sinks is eased by materials recovery as the resources are retained within the economy rather than discarded. However, recycling, composting and incineration technologies each have their own knock-on environmental impacts, which counterbalance and in some cases outweigh the benefits of recovery.

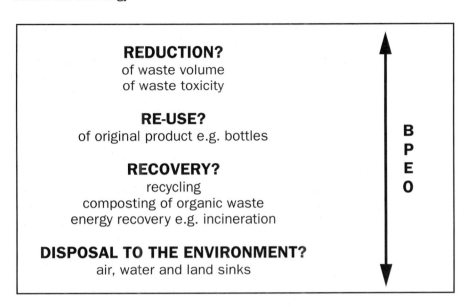

Figure 4.3 *The waste management hierarchy as a decision-making framework*

For example, whilst most analysts agree the recycling of aluminium drink containers is highly desirable in environmental terms, paper recycling is more problematic. Refining aluminium from bauxite ore requires very large amounts of energy compared to the energy required to collect, sort, transport and reprocess used drink cans. For the paper industry, the difference in energy and water use between processing primary and secondary pulp is much smaller. If recycling entails the transportation of waste paper over long distances prior to reprocessing the increase in transport related fuel use and exhaust emissions may outweigh any environmental benefit over the process as a whole.

Reuse of the product itself can avoid the disadvantages of recovery technologies. Extending the useful life of products by designing them for longevity – robustness in use and straightforward repairs – is simple in theory but runs counter to the postmodern trends toward consumerism and the throwaway society discussed in Chapter 2. It is also more difficult in a globalised economy where products are used thousands of miles from where they were manufactured. Because of the pollution caused by transportation, reusing even quite simple products such as returnable glass drink containers does not make environmental sense without local bottling plants and small-scale distribution networks. Reuse of older consumer goods and industrial plant which consume energy and resources is not always a good option environmentally if more up-to-date designs are more efficient or have better pollution control characteristics.

Minimisation is the highest tier of the hierarchy and avoids the disadvantages of all the lower stages by not producing the waste in the first place. Increased resource and energy efficiency in manufacturing processes is one example of minimisation; another is the so-called light-weighting of packaging so that the minimum mass of materials consistent with the effective labelling and protection of the product is used.

The hierarchy is sometimes criticised for being too rigid and prescriptive but this is to misunderstand its basis, which is as a framework for decision making, not a set of prescriptions to be followed in all circumstances. In some circumstances choosing lower tier options can result in a better outcome environmentally and economically. A good but contentious example is the recycling (reprocessing) of spent nuclear fuel – a process which is condemned by environmental groups as expensive and polluting (Chudleigh and Cannell 1984). For any material, recycling, reuse and minimisation will only make sense if the combined economic, environmental and social costs can thereby be minimised – that is if the BPEO can be achieved.

Some radical approaches to production suggest that substantial increases in the resource efficiency of developed countries' economies are technological possible. Factor Four (von Weizsäcker, E. *et al*. 1997) and Factor Ten (Factor Ten Club 1994) are analyses proposing four or ten fold increases in resource productivity as feasible. Such increases would depend on a realignment of the relationships between producers, consumers and products. The conventional conception of this relationship is that producers manufacture the product which is sold to the consumer, who then disposes of it at the end of its useful life. However, in many instances, it is the service that the product performs which is desired by the customer, not the product *per se*.

Consumers don't want electricity, or petrol, or gas – they want light, heat and motive power. Stretching the concept of environmental services, developed in Chapter 1, from the product to the service it performs for the consumer can lead to a more creative approach to getting maximum resource productivity with minimum waste. When producers shift from selling their product to providing services, they have incentives to minimise economic and environmental costs over the whole product life cycle. Using techniques called demand management, they work with consumers of their products to maximise resource efficiencies and minimise wastes.

A utility that sells electricity will try to increase its profits by generating and selling more units each year. A company selling energy services (i.e. light and heat) would have a vested interest in providing these services using the smallest amount if energy possible. As well as providing fuels such as electricity and gas to the consumer they would also provide energy efficient appliances (light bulbs, boilers) and insulation for the building fabric. The price charged for each unit of delivered energy would rise, but consumers would be happy because their overall bills would fall, fewer units being needed to provide the same level of service.

The same approach has been successfully used with water. Fitting water meters for domestic customers can be criticised as regressive – poor consumers may suffer hardship if they cannot afford the charges. In Los Angeles, however, the introduction of metering coincided with a scheme to replace showerheads and toilets (and light bulbs) in disadvantaged neighbourhoods, resulting in savings of $30–$120 per household on utility bills (Hawken *et al.* 1999).

Moving to a service approach is not just an option for utilities. Xerox is a good example of a company that has adopted this approach, and claims savings of $35 million as a result in 1996 alone (Xerox 1998). On the basis that its customers want photocopies, not photocopiers, it leases rather than sells its machines. At the end of the lease period the machine is returned to the company for reconditioning of either the whole machine or its components. These are then used on the production line in preference to newly manufactured parts.

Extending this approach to transport, however, demonstrates some limitations. It could be argued that what people want is not mobility (as provided by trains, boats and planes) but access – to shopping, employment, health-care, education and leisure opportunities. Therefore, a decent public transport system is an excellent alternative to a private system based on cars, congestion and pollution. Opportunities offered through telecommunication are equivalent to those requiring travel and face-to-face contact. But, even if the quality of internet shopping, health care etc. is as good as that offered in shops and health centres, in many instances mobility for its own sake is exactly what people desire. In addition, they much prefer that mobility in automobiles rather than buses. Not for the first time in this book, consumer desires confound environmental policy makers.

Conclusion

Policy makers need scientific information when defining environmental problems and seeking solutions to them. However, the nature of the scientific method means that the meaning of such information will be different to different actors within the policy process. When scientific information is partial or uncertain no regrets strategies and/or the precautionary principle can provide frameworks for policy making.

There are deeper problems, however, with the status of scientific knowledge within the policy process. The privileging of scientific 'objective' knowledge above other kinds of knowledge can be challenged on both philosophical and political grounds. In postmodern cultures the role of the 'expert' in the policy process is problematic, especially if expertise is used to exclude non-expert participation in policy making.

Technology will inevitably play a vital role in any strategy to meet human needs in a sustainable way. However, given the potential that technology has to wreak damage to the environment and to social welfare, the 'appropriateness' of technology is a fundamental issue. Preventative environmental management techniques are preferable to end-of-pipe approaches. Ecological modernisation builds on such preventative approaches to propose a development route offering simultaneous economic and environmental advantages, provided the necessary infrastructural changes can be achieved.

Further reading

An environmentalist's critique of science and the scientific method can be found in Porritt (2000). For a review of the academic literature see Pepper (1996), especially chapters 3 and 5. Harremoës et al. (2002) charts the development of the application of the precautionary principle in environmental policy through case studies. Schumacher (1973) is still the classic text on appropriate technology. Douthwaite (2002) considers how best to manage technological developments to meet human needs in the twenty-first century. For a well-written and accessible account of the theory and practice of preventative environmental management see Jackson (1996). Hawken et al. (1999) explains Factor Four and demand management, whilst a good introduction to ecological modernisation can be found in Gouldson and Murphy (1997). A more detailed account is Mol and Sonnenfeld (2000), which provides an overview of the theory of ecological modernisation, including its applicability in developing countries.

5 Corporate environmental policy making

- The drivers which are shaping environmental policy development in businesses
- Environmental management systems and other corporate responses to these drivers
- The relationship between corporate environmental management and sustainable development

Corporate environmental policy: the context

Limited liability for firms underpins the capitalist system by encouraging risk-taking and entrepreneurship. By setting up as a public limited company (in the United Kingdom) or corporation (in for example the United States) firms can attract capital from investors who know that, although their investment funds are at risk, they will not be responsible beyond this for the debts of the company. Similar protection is offered to directors and employees of the firm, provided they operate within the law and in good faith. If the company fails, its debts fall upon its unfortunate creditors, once all assets have been sold. Limited liability is undoubtedly a privilege. Society is rewarded for its grant of limited liability to firms by the wealth that is created through the resulting enterprise.

Firms and their shareholders benefit directly. For their investment the shareholders in a company expect returns, usually in the form of dividends and rising share value. Thus the primary function of a firm is to generate value for shareholders by trading for the maximum profit it is able to achieve. On the face of it this provides a strong incentive to behave ruthlessly with little regard for social or environmental responsibility. However, businesses have always had to adapt their practices to keep up with developments in legislation and social attitudes. The choice has often turned out to be one of timing only: whether to be in the vanguard or the rear of the changes. During the period of industrialisation in the eighteenth and nineteenth centuries a clearly differentiated pattern of response by businesses is evident. On issues such as the abolition of slavery and the elimination of child labour a minority was quick to detect impending change in the social climate. In some cases they led the campaign for change, leading by example in their own business practice, for example the chocolate manufacturing business owned by the Cadbury family in Birmingham, England. Other companies, however, fought against these trends, using their political and economic power to put off the day when they would reluctantly have to comply with new laws banning the practices upon which their profits were dependent.

In the twentieth century the business agenda has continued to adapt, although the issues have moved on. In a gradual process over several decades, the growing influence of trade unions led to improvements in health and safety provision in the workplace. In the 1980s quality management became an important component of business competition as consumers became more demanding and discriminating. The mix of pro-active and reactionary responses from businesses has also been evident in the emergence of environmental issues on to the business agenda, as Boxes 5.1 and 5.2 illustrate.

As working practices changed, legislation and regulatory requirements became ever tighter in a cycle of continuing improvement. According to Porritt (1997) this process can be categorised into three broad phases. At first, during the 1960s and 1970s, there was from business an outright denial that any significant problem existed and therefore confrontation between environmental groups on the one hand and business and government on the other. In the second phase, government began to respond to pressure from the environmental movement by tightening the regulation of business, which responded reactively, not least because of complementary pressure from consumers. In the third phase, dating from about the 1990s, some businesses emerge as proactive partners in this process, seeking to do more than is legally required, in order to achieve their own sustainable economic development. This is the process of ecological modernisation, introduced in the last chapter.

Where preventative environmental management based on 'no regrets' strategies offers short, medium or even long term economic benefits, the motivation for businesses to adopt these approaches is clear. However, there are other drivers at work. Five broad categories of stakeholders can be identified, each with the potential to push for improved corporate environmental management (Howes et al. 1997; Nelson et al. 2001). These are:

- government, through changes in legislation and regulation;
- customer preferences or demand;
- local communities and non-governmental organisations (NGOs);

- investors;
- employees.

However, sustainable development entails much more than just reduced environmental impact. An approach based on the 'triple bottom line' has been developed to allow companies to engage with the whole range of sustainable development issues:

> The triple bottom line (TBL) focuses corporations not just on the economic value they add, but also on the environmental and social value they add – and destroy. At its narrowest, the term triple bottom line is used as a framework for measuring and reporting corporate performance against economic, social and environmental parameters. At its broadest, the term is used to capture the whole set of values, issues and processes that companies must address in order to minimise any harm resulting from their activities and to create economic, social and environmental value.
>
> (Nelson *et al.* 2001: 6)

Government regulation

Responsible businesses have always sought to comply with the law and failure to do so can have disastrous consequences for the balance sheet (Box 5.1). Even minor breaches can damage profits beyond the cost of fines imposed. If the company is liable to pay compensation for damage caused it may be able to meet the costs of reparation from insurance. However, future premiums will be very high or, in the worst cases, cover refused. For companies that can demonstrate sound environmental policies which reduce the risk of incidents and prosecution insurance premiums may be reduced.

The very rapid development of environmental legislation over the last three decades, however, has meant that anticipation of forthcoming legislation can be as significant a driver as compliance to existing laws (Box 5.2). Anticipation might result in investment to meet new standards before they are enforced. Alternatively the company may lobby policy makers to lessen the impact of the proposed legislation on their industry. The structure and character of such lobbying are analysed in the next chapter.

Box 5.1

Asbestos: Turner & Newall

Asbestos used to be thought an ideal material for many applications such as heat and fireproofing of buildings. However, the airborne fibres are a severe health hazard and symptoms of asbestos-related disease can take more than thirty years to develop. Asbestos fibres can cause:

- asbestosis, an incurable lung disease;
- lung cancer;
- mesothelioma, a malignancy of the lung lining or abdomen.

In the United Kingdom an estimated 3,000 people a year die of asbestos-related disease (Tweedale 2000: 275). For men born in the 1940s the eventual death rate from mesothelioma is predicted to reach one in a hundred.

The large-scale production of asbestos based materials was pioneered by the British firm Turner Brothers (later Turner & Newall (T&N)) around 1880. Profits were consistently good and the company grew rapidly so that by the Second World War it had operations in Canada, India and several African colonies.

Although concern about health of workers in British asbestos factories was first expressed in 1898 by the Chief Inspector of Factories and Workshops, it was not until 1931 that regulations forced T&N and other asbestos manufacturers for the first time to take measures to control dust within the workplace. This was in response to a government-commissioned report, which found that 80 per cent of asbestos factory workers with more than twenty years' exposure to the mineral had asbestosis. Monitoring of the health of those who worked in the dustiest areas was introduced, together with a scheme for compensating those so badly affected by asbestosis they could no longer work and their dependants once they died.

While ending the most dangerous practices and requiring safer working practices such as ventilation, these changes did not make asbestos factories safe. The regulations were not always followed correctly, and enforcement was weak and ineffectual. When claims for compensation were made the usual response of the company was to deny liability. Inquests were routinely attended by medical and legal representatives of the company, who did all they could to persuade the coroner and jury that factors other than asbestosis had caused the death.

In 1955 the link between asbestos exposure and lung cancer was confirmed by an epidemiological study initiated by T&N and based on the post-mortem reports of 115 of their own workers. The company attempted to suppress publication. T&N then took a leading role in the establishment of the Asbestosis Research Council (ARC), in partnership with other asbestos producers. Thus the industry hoped to control the research agenda and maintain a veto over the publication of ARC-funded studies.

But scientific evidence of the dangers of asbestos continued to accumulate. In 1965 an independent study into the occupational and residential histories of mesothelioma victims in London discovered that stores and office staff in the asbestos industry were

continued

Box 5.1 continued

also among the victims, as were those sharing the homes of asbestos workers. Even living within half a mile of an asbestos works appeared to be a significant risk factor.

By this time the company was preparing for an asbestos ban by diversifying into alternative products. However, liability claims started to accelerate, fuelled by the growing literacy and assertiveness of employees and the support of trade unions or state-funded legal aid. Although liability was borne by the company's insurers, T&N itself was responsible once the insurance limit was reached.

The most significant litigant was Chase Manhattan Bank, which sued T&N for $185 million – the cost of stripping asbestos from its Wall Street headquarters. Although the case was eventually decided in T&N's favour it resulted in the opening to public scrutiny of the company's extensive archives and much ammunition for other claimants.

The company's profits were being swallowed up by escalating claims and in 1997 it was taken over by the US-based Federal Mogul in a rescue bid, but the growing liability bill was too much. In October 2001 Federal Mogul was itself forced into Chapter 11 bankruptcy. At the time of writing T&N is still trading under protection from its creditors and recent UK judgements have been in its favour, demonstrating the difficulty of proving legal liability between exposure to toxins with long latency periods and long-term chronic diseases.

References: Jeremy (1995); Newhouse and Thompson (1965); Tweedale (2000).

Discussion points

1   Identify the stakeholder groups in the case study above. Which groups benefited in the short term from the company strategy up to about 1960? Which groups benefited in the long term from this strategy?

2   What lessons about the links between social, environmental and economic sustainability can be drawn from the case study?

Box 5.2

Du Pont and the CFC phase-out

Chlorofluorocarbons (CFCs) are a class of chemicals which were first developed in the 1930s by the Du Pont corporation. They have been very useful industrially as they are chemically stable, low in toxicity and non-flammable. CFCs work well as heat transfer agents in refrigeration and air conditioning systems. Other uses are as aerosol propellants, in the manufacture of foams, and as solvents in the electronics industry.

Box 5.2 continued

Although Du Pont's patents on CFCs had expired by the 1950s it remained the world's largest manufacturer of the chemicals. By the mid-1980s Du Pont produced just over 20 per cent of world demand and this activity was responsible for 2 per cent of corporation profits. However, in March 1988 the corporation decided to abandon past investment in manufacturing capacity and to phase out the production of these chemicals, even though no good substitutes were available for many of the uses of CFCs. This case study examines the reasons for the decision.

Worries about the environmental safety of CFCs were first expressed in the 1970s when two American scientists, Rowland and Molina, proposed the theory that CFCs in the upper atmosphere (stratosphere) could undergo chemical breakdown to produce free chlorine atoms. It was suggested that these could then catalyse a chain reaction, resulting in the conversion of ozone (O_3) to oxygen molecules (O_2). Each free chlorine atom could catalyse many thousands of these reactions (see diagram). Stratospheric ozone plays a vital role in preventing most of the short-wave radiation from the Sun (UVB radiation) from reaching the Earth's surface. UVB is known to cause adverse health effects in humans (e.g. skin cancer and cataracts) and damage to terrestrial and marine eco-systems.

The theory that CFCs might cause stratospheric ozone depletion was therefore alarming – but there was no evidence to support it. Stratospheric ozone levels were subject to considerable natural variability and difficult to measure. Considerable research would be needed to confirm or deny the hypothesis and Du Pont contributed by helping to establish a panel of CFC producers who would co-operate in funding an atmospheric science research programme. The corporation also allocated research funds to programmes to identify potential substitutes for CFCs.

There were calls for a ban on CFCs which the corporation resisted, claiming that denying industry and consumers the benefits of CFCs could not be justified in the absence of scientific proof of damage. However, as early as 1974 the corporation promised to stop production if and when such proof emerged. Consumer pressure in the mid-1970s forced legislation to ban CFCs from aerosols in the United States, Canada, Norway and Sweden but other uses of the chemical were unaffected and these markets continued to grow, as did aerosol markets in most of Europe and the developing world. Despite the bans, by the late 1980s global production of CFCs had attained again the peak levels seen fifteen years previously.

During this time atmospheric models had been developed to predict the effects of CFCs on ozone and the results were reassuring. Provided CFC use did not expand at too fast a rate the damage to the ozone layer was predicted to be small. Satellite monitoring of stratospheric ozone by NASA showed no evidence of deterioration.

So, in 1985 it was a considerable surprise when a British research team in Antarctica published evidence that stratospheric ozone over the South Pole was 40 per cent thinner than usual during the spring months. NASA rechecked their records and found that, although they had gathered similar data, the relevant readings were so low that they had been rejected by their computer's automatic error-detecting software. The decline had been occurring for a decade and proved the predictions generated by the atmospheric models which had been developed to be too optimistic.

continued

Box 5.2 continued

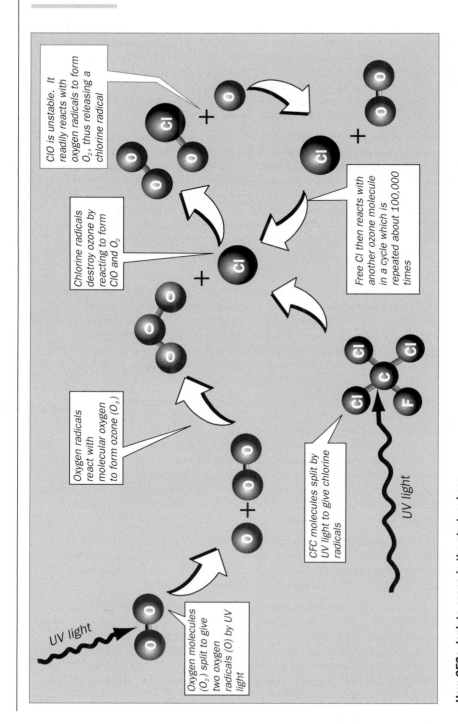

How CFCs deplete ozone in the stratosphere

Box 5.2 continued

The response of Du Pont was to announce the phase-out of CFC production, the first corporation to do so. What were its reasons? Given the emerging scientific consensus that the ozone hole was caused by CFCs, the corporation would have been subject to campaigns and adverse publicity had it not taken action, especially given its 1974 promise to do so. In addition, by 1988 international negotiations for a global production ban through the Montreal Protocol were well advanced and to some extent the corporation was anticipating the inevitable.

However, there were strategic advantages in being ahead of the pack and joining the coalition of interest groups calling for a ban. Developing CFC substitutes and building markets for them would be very capital-intensive, but this was the path Du Pont was choosing. Furthermore, CFCs were out of patent, but if Du Pont was first to market with substitutes it would gain exclusive benefits from its patents on these. However, unless CFCs were banned throughout the world the markets for these substitutes would be difficult to establish as consumers would continue to choose the cheaper CFCs. The corporation's pro-active environmental policy and lobbying therefore had two advantages. The corporation could promote itself as corporately responsible whilst gaining competitive advantage for the future.

References: Reinhardt (1992); UNEP (2000a).

Discussion points

1 Was the US ban on CFCs in aerosols an example of a no-regrets policy or of the precautionary principle (Chapter 4)?

2 What would have been the consequences for Du Pont if it had stopped manufacturing CFC unilaterally at the time of the US aerosol ban?

Customers

Companies sell their products and services either to individual members of the public, or to other companies or organisations, or to both. Publics in the developed world began to demonstrate what has been called 'green consumerism' (Elkington and Hailes 1988) in the late 1980s at a time of increased anxiety about the environment (Grove-White 1993). The prospect of a new niche market led to a proliferation of products aimed at green consumers. Some of these had dubious environmental credentials and the boom was short-lived. However, a lasting effect has been heightened awareness amongst a sizeable minority of consumers of the environmental and social issues associated with certain goods, such as detergents, coffee and paper products, although this is rarely based on specific knowledge.

More recently consumer attention has shifted from the pros and cons of specific products to the reputation of companies and the brands that they own. Scandals and controversies, whether environmental or social, can damage brands and sales. Shell discovered this when the company tried to tow the Brent Spar oil platform to the deep ocean for dumping in 1995. Sales from Shell petrol stations fell across Europe and there were even firebomb attacks on the company's premises in Germany (Elkington 2001a). The long-running boycott of Nestlé by those opposed to the marketing of breast milk substitutes in developing countries is another example.

Corporate customers can be more demanding. For some time it has been common for large companies, government departments and local authorities to require some measure of environmental performance as part of their procurement policies (Hopfenbeck 1993). At one extreme this may be merely a check that the supplied product meets certain environmental criteria. At the other, there may be a requirement that all suppliers adopt a certified environmental management system (see below), with perhaps the offer of help and advice on how to achieve this.

This trend is having a large impact in some industries, such as motor components, where small and medium-size enterprises (SMEs) supply multinational corporations such as Ford and BMW. Larger corporations tend to be in the vanguard of environmental policy initiatives because they have the resources to research and implement the necessary changes and the incentive of a valuable brand to protect from the risk of adverse publicity or to enhance through a pro-active and responsible stance. SMEs are less likely to have taken action beyond that which is legally essential, unless pressure from their corporate customers has forced them to.

Local communities and NGOs

Companies whose operations cause nuisance or potential hazards to communities near their sites will find it advantageous to adopt the triple bottom line approach, which includes annual reporting on economic, social and environmental performance. Transparency and a willingness to engage with outside groups can build trust and pre-empt adverse publicity. This is true also of non-local issues, including companies' activities in the developing world, such as the use of child labour or payment of poverty wages. Since the Brent Spar incident Shell has adopted this approach, using environmental and social NGOs to facilitate dialogue on the company's performance with actual and potential critics in order to shape its policies (Murphy and Bendell 2001).

Investors

Most small shareholders are content to allow economic issues only to guide their investment, although there is a grand tradition of activists buying small shareholdings in controversial companies, such as Rio Tinto-Zinc, in order to attend annual meetings and protest. Institutional investors, such as venture capital firms and pension funds, have begun to demonstrate a more interventionist approach, demanding changes for companies if it is perceived that these issues are being neglected (Nelson *et al*. 2001). Indeed, UK pension funds are now required to report annually on their investments, using a triple bottom line approach. Because poor social and environmental performance can damage profits, investors are wary, not only of companies with a poor track record on these issues, but also of those where the risk of future poor performance has not been minimised by, for example, the adoption of environmental management systems.

In the United Kingdom there are an increasing number of Socially Responsible Investment (SRI) funds whose investments are chosen using social and/or environmental criteria (Millar 2001). Only companies whose activities meet these criteria will be eligible for this source of capital. Some funds have stricter criteria than others. At one extreme are the so-called negative screeners, funds that merely seek to avoid the most controversial investment areas (armaments, tobacco etc.). In the middle ground are the positive screeners who seek to invest only in firms making a positive contribution to environmental and social capital. The most pro-active funds go further than this, engaging with companies to encourage and assist in continued improvements in environmental and social performance.

Employees

Companies are increasingly recognising the importance of environmental and social issues to the morale and motivation of their employees. Brand image is important when recruiting staff. Applicants will avoid companies with a poor social or environmental record. Existing staff are able to judge the sincerity of environmental and social policies much more easily than customers, investors or other external stakeholders. They will be discouraged by poor performance or, alternatively, motivated when this improves, especially if they are involved, through, for example, staff development programmes in environmental management or social volunteering initiatives.

Corporate environmental policy in action

Since the early 1970s, these pressures have led to rapid developments in the formulation and implementation of environmental policies at the corporate level. The first response was the development of the environmental audit as a tool to measure the impacts of organisations. Next, environmental management systems emerged as a way to actively monitor, control and reduce environmental impacts, not just of the organisation, but also of its suppliers and the end users of its products. Systems to measure and control sustainability impacts are now being developed as the next logical stage of this process.

Environmental auditing

Every company undergoes each year a financial audit, a process by which the flows of money in and out of company's accounts are tracked over a year in order to produce a profit and loss account. An independent accountant, who identifies any problems and brings them to the attention of the board and the shareholders, does this in a systematic way. These problems may be with the money flows themselves if funds are missing. More commonly auditors will flag up problems with the financial systems within the firm, or technical breaches of financial regulations.

Environmental auditing adopts much the same approach. The audit can be carried out as an internal exercise by and for management but will have more credibility if performed by an independent auditor. The scope of the audit needs to be clearly defined. Audits may cover:

- the direct impacts of the firm's activities, either as a whole or for one or more issue – energy, water and other resources used; emissions and wastes produced; noise and other nuisances such as heavy transport movements;
- the impacts created by the firm's suppliers – for some businesses such as retailers supply chain impacts are likely to be much more significant than the operation of the core business;
- the impacts created by one or more product when in use and at the end of its life cycle (perhaps assessing the cradle-to-grave impact using life-cycle assessment (LCA));
- the extent of the firm's compliance with relevant legislation;
- the robustness of any environmental management systems in place, including the extent of compliance with any environmental policies and targets previously adopted;
- the entire firm or just one operational site.

Corporate environmental policy

To be useful, of course, audits must contain clear recommendations for action where appropriate. These are usually put into effect by the development of an environmental policy and its implementation by an environmental management system. The quality management systems developed during the 1980s provided the procedural basis for the environmental management systems that emerged in the 1990s.

Because environmental management systems exist in order to implement environmental policies, the formulation and review of such policies lie at the heart of any EMS (Figure 5.1). In the Introduction to this book environmental policy was defined as 'a set of principles and intentions used to guide decision-making about human management of environmental capital and environmental services'. At the corporate level an environmental policy should include within its scope:

- an explicit statement of the principles upon which it is based;
- a commitment to identify and comply with all relevant legislative requirements;
- statements explaining how the information needs of stakeholders (e.g. employees, customers, suppliers, regulators, communities) are to be met through record keeping, communication and training;

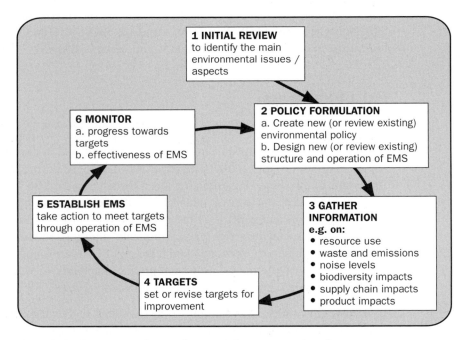

Figure 5.1 *The stages of an environmental management system*

- statements explaining the company's expectations, in terms of environmental standards, from its suppliers, the companies it supplies and those selling its products or services;
- commitments on specific issues, such resource use, resource efficiency, waste and emissions; product life-cycle impact etc.;
- a commitment to continually monitor the environmental impact of the company's activities;
- a commitment to continuous environmental improvement;
- a commitment to review and update the policy at a regular intervals (but no minimum frequency is necessarily specified).

It is essential that environmental policies are formulated and endorsed by the top management of a company and that their role and purpose are communicated throughout the organisation. If these fundamental requirements are not met the policy is unlikely to be implemented successfully, if at all.

Any system by which a firm develops an environmental policy, sets targets for improvement and monitors progress towards these targets through auditing, in a continuous loop, is an environmental management system (Figure 5.1). If it is purely an in-house system, however, it is of little use in convincing external stakeholders of the company's commitment to environmental improvement. Two certified environmental management systems have been developed to meet the need for external recognition and independent verification. These are the International Standards Organisation (ISO) series of standards ISO 14000 and the European Union Eco-management and Audit Scheme (EMAS).

ISO 14000 series

The ISO 14000 series is available to companies across the world and has been adopted in the European Union as EN ISO 14000 and in the United Kingdom as BS EN ISO 14000. The series was developed during the 1990s and, in the United Kingdom, replaced British Standard BS7750. The aim of the series is to cover 'standards in the field of environmental management tools and systems' (Hortensius and Barthel 1997: 21). Thus there are standards in the series on environmental auditing, environmental performance evaluation, environmental labelling, LCA, terms and definitions, and environmental aspects of product standards; as well as the ISO 14001 standard on environmental management systems.

To achieve ISO 14001 an organisation must develop, to the standard, an environmental management system with the following stages. These have been numbered to correspond with Figure 5.1. The stages are:

- identification of environmental aspects (1 and 3);
- the development an environmental policy (2);
- establishment of relevant legal and regulatory requirements (3);
- development of environmental objectives and targets (4);
- the establishment and maintenance of an environmental programme in order to achieve its objectives and targets (5);
- implementation of an EMS, including training, documentation, operational control, and emergency preparedness and response (5);
- monitoring and measurement of operational activities, including record keeping (6);
- EMS audit procedures (6);
- management review of an EMS to determine its continuing suitability, adequacy and effectiveness (2).

(Hortensius and Barthel 1997)

Certifiers are independent consultants, recognised as competent by a national standards body such as the British Standards Institute (BSI). They are engaged to check that the system conforms to ISO 14001. Documentation is key to the certification process and is likely to be voluminous. This is an important disincentive for SMEs, which often lack the administrative capacity or will to assemble and maintain such records. The certifiers will check paper or computer records of the operation of the system and certification will only be granted once the checks have confirmed the system is operating as intended and in conformity with the standard.

ISO 14001 has proved successful to the extent that it has been adopted by a significant minority of organisations, and especially by large and multinational companies. However, it has its critics. Gleckman and Krut (1997) go so far as to claim that the adoption worldwide of ISO 14001 has led to a reversal of previous trends towards improved corporate environmental performance because companies now aspire to meet only the minimum requirements of the standards.

The main problem is that no external targets or benchmarks for environmental performance are set by the standard, beyond compliance with relevant legislation. ISO 14001 has been designed so that organisations seeking certification set their own performance measures and targets, and the certifier checks conformance with these. The more conservative the target, the easier it will be to achieve compliance and therefore certification. Thus, not only is there no incentive to go further than minimum legal requirements, there is a built-in disincentive to the adoption of challenging targets.

A further criticism is the use of the term 'environmental aspects' (meaning energy and resource inputs and waste outputs of production systems and products) rather than 'environmental impacts'. Impacts would imply a quantified assessment of the actual effects of resource extraction and waste production on the environment, a complex task especially for SMEs. Inputs and outputs are much easier to track through normal business processes such as invoicing for the purchase of raw materials and waste disposal.

EMAS

EMAS was introduced by the European Union in 1993. Although the standard passed into law as a regulation (EEC 1836/93), participation is voluntary. The EMAS and ISO 14000 series are compatible, so that it is possible to gain accreditation to both simultaneously, or, alternatively, to gain first one standard and then the other. However, there are significant differences between ISO 14000 and EMAS, and the Comité Européen de Normalisation (CEN) has developed a bridging document specifying the steps needed for certified holders of ISO14001 to gain EMAS (BSI 1997). In summary, the main points of difference are:

- ISO 14001 applies to an organisation, EMAS to a site.
- ISO 14001 is 'certified' and can rely on company self-assessment to a greater extent than EMAS, which is 'verified', a process which entails a greater depth of investigation by the independent assessor.
- ISO 14001 requires only limited public disclosure; for EMAS accreditation a verified environmental statement must be issued into the public domain.
- Whilst both standards include a commitment to continued improvement in environmental performance there is no external reference for this performance under ISO 14001; EMAS requires a commitment to work towards BATNEEC.
- ISO 14001 does not specify a maximum time period for the EMS cycle; EMAS requires a maximum period of three years.

EMAS was developed for companies within the industrial sector. A variant was subsequently developed for local authorities in the United Kingdom (CAG 1993). This was based not on sites but on operational units within local authorities, although with a requirement that the authority as a whole takes responsibility for corporate overview and co-ordination of environmental management. The focus is not only on direct environmental impacts but also on indirect effects through service delivery. This approach recognises that the environmental impact of decisions emerging from planning departments will be of vastly more significance than the waste paper in their recycling bins.

Managing sustainability

As leading-edge companies have moved from a focus on environmental management to a triple bottom line approach, interest has grown in developing an externally verified set of sustainability standards. The first stage of this will be incorporating social issues into the existing frameworks for financial and environmental accounting. Some schemes have been developed which do just this, such as the Global Reporting Initiative (GRI), designed by the Coalition for Environmentally Responsible Economies (CERES) in partnership with the United Nations Environment Programme. This shares with ISO 14001 a process approach – there is no provision for external referencing. However, companies that adopt the GRI approach have the opportunity to address environmental and social issues by producing a report, which includes:

- Statement by Chief Executive Officer.
- Profile of Reporting Organisation.
- Executive Summary and Key Indicators.
- Vision and Strategy.
- Policies, Organisation and Management Systems.
- Performance.

(Accountability *et al.* 2001)

In the United Kingdom the government-funded Sigma project has built on the GRI approach to develop a set of sustainability guidelines for business (Table 5.1). These provide a management framework for sustainability issues. It can be seen that Sigma seeks to build on and be compatible with existing quality and environmental management systems by adopting the same framework of policy formulation, target setting for improvement, and monitoring of progress through a looped audit cycle.

However, the issue of external referencing of performance is as fundamental to sustainability management as it is to environmental management. ISO 140001 has no external reference point, as we have seen, and EMAS uses BATNEEC. Compliance with both standards will result in improved environmental performance to some extent. However, as was seen in Chapter 3, sustainability as a goal of environmental policy is a much more demanding criterion than BATNEEC and similar targets. Strong sustainability would demand that the organisation demonstrate that its activities do not deplete, or actually enhance, environmental capital. An equivalent criterion for economic capital is of course unproblematic – annual financial reporting systems provide the evidence on this. But for social sustainability the same problem arises. Although there are requirements under both GRI and Sigma to engage with, consult with and report to relevant stakeholders, companies using these systems may begin

Table 5.1 *The SIGMA management framework: ten key phases*

Management phase	Phase purpose
Sensitisation and awareness	To secure sufficient commitment to undertake a baseline review and to integrate sustainability and stakeholder engagement into core processes and decision making
Baseline review	To establish the organisation's values, strategies and performance with regard to sustainability
Actions, impacts and outcomes	To understand and manage the relation between organisational actions, impacts and outcomes
Legal, regulatory and other requirements	To understand and manage current and future legal and self-regulatory requirements
Strategic planning	To formulate long-term sustainability strategies
Tactical planning	To develop a series of tactical plans that address organisational strategies and the impacts and outcomes identified by the organisation
Communication and training	To align internal and external communications and training with strategic and tactical planning
Control and influence	To ensure that actions, impacts and outcomes work in alignment with and support tactical and strategic planning
Monitoring, objective evidence, feedback	To generate and maintain efficient internal and external feedback loops to monitor progress against stated values, strategies, performance objectives and targets
Reporting progress, tactical and strategic review	To meet the information needs of internal and external stakeholders and incorporate feedback into effective strategic and tactical reviews culminating in appropriate change

Source: BSI (2001)

the journey towards social sustainability but are unlikely to arrive, not least because the destination is undefined.

The Natural Step

The Natural Step (TNS) is an approach to sustainability in organisations that seeks to provide an overarching reference framework for environmental and social sustainability. TNS emerged in Sweden in the late 1980s and is now active in Australia, the United Kingdom and the United States. TNS is based on the idea of four 'system conditions' which are claimed to be both necessary and sufficient for sustainability. These are set out in Table 5.2. The first three were developed from basic scientific principles, notably the second law of thermodynamics. They set out the environmental limits on human activity. The fourth condition links human needs to environmental limits by prescribing equity and efficiency in the use of environmental services to meet human needs. If all four conditions are met neither environmental nor social capital will be depleted.

The simplicity of the TNS conditions is at once a strength and a weakness of the concept. Although superficially easy to understand in general terms, applying and measuring the application of the system conditions in

Table 5.2 *The Natural Step's four system conditions*

In order for society to be sustainable, nature's functions and diversity are not systematically:

- **subject to increasing concentrations of substances extracted from the Earth's crust** This means that carbon from fossil fuels, minerals and metals should not be increasing in the biosphere. Substitution and increased efficiency of use, including recycling, are required to slow down the rate of flow through the resource cycle towards the sustainable rate

- **subject to increasing concentrations of substances produced by society** Synthetic chemicals which are persistent (e.g. DDT (Box 1.1) and CFCs (Box 5.2)) should not be emitted to the biosphere. Labile substitutes, used efficiently, will avoid these problems

- **impoverished by physical displacement, over harvesting or other forms of ecosystem manipulation** Environmental services must not be over-used, eco-systems must be carefully managed, and a conservative approach taken to anthropogenic environmental change

- **resources are used fairly and efficiently in order to meet human needs globally** This is the equity condition – ensuring fairness in the distribution of environmental services and other forms of wealth. For the developed world this means frugality and efficiency as well as regard for future generations

Source: The Natural Step (2002)

any given situation can be difficult. Despite this several large companies such as IKEA, the Swedish furniture retailer, Yorkshire Water and the Co-operative Bank in the United Kingdom, as well as multinational companies such as Air BP, which supplies a tenth of the world's aviation fuel, have adopted the Natural Step approach by becoming TNS Partners.

The first step to adopting TNS is 'backcasting' – developing a vision of the characteristics that a particular business would have in the future if it were to achieve compliance with the four conditions. The challenge then is to develop strategies to reach this goal. Some companies, by the very nature of their business, find it easier to apply the TNS conditions than others. Service companies, such as Yorkshire Bank, have made large strides by developing credit and cash cards made from non-PVC plastic; reducing internal energy consumption; adopting transport, paper and IT management policies which minimise the consumption of environmental services; and offering preferential loans to other TNS partners. Some manufacturing and retail companies can find it relatively easy as well.

Interface, a US TNS partner in the carpet business, has developed an approach based on leasing carpet tiles to customers rather than selling them carpets. It thus retains responsibility for its product throughout its life cycle. Damage or wear to a small area of flooring means replacement of the affected tiles only, not the whole carpet. By investigating renewable and recycled raw materials for the manufacturing process, and recycling options for discarded tiles, the company is opening up opportunities for greater compliance with the system conditions.

A company whose business is fundamentally wasteful in its use of environmental services will obviously find it correspondingly difficult to change in line with system conditions. Air BP is a good example here as its core business at present contravenes all four conditions. It depends on the extraction of large quantities of fossil fuel; the wastes from combustion of this fuel are accumulating in the atmosphere; this build-up may be causing global warming and knock-on effects on natural systems; and the end purpose of this activity is air travel, which is far more of a luxury than a basic human need. The company's involvement is explained thus:

> The reason, says Vivienne Cox, the company's Chief Executive Officer, is that TNS is a tool you can actually use: 'I respect deep green views and I recognise the wider shift in consumer opinion in favour of clean fuels which do no damage to the natural environment. We believe we can be a leader in providing fuels of that quality, and TNS is a helpful, pragmatic approach which recognises the business

realities without compromising the objective.' At the same time 'the first three system conditions make sense to me as a scientist. For us, fuel quality is a fundamental service we provide our customers and I'm determined that Air BP should set the standard for the whole industry.' She is also convinced that the products which make more sense environmentally will have a competitive advantage.

(Henderson 1998)

This illustrates well a common criticism of TNS – that although it gives industry a set of conditions to aspire to, and evidently some companies are engaging enthusiastically with these, it sets no time scale for the attainment of the conditions. Moving towards sustainability is important, but so is the pace at which this is happening. Although early steps that businesses take towards sustainability may bring increased efficiencies, and therefore greater profits, later decisions may be more painful and costly.

The radical critique and some conclusions

ISO 14000 assures continuing improvement but against no performance benchmark standard beyond the *status quo* and at a pace chosen by the company. EMAS is externally referenced, but to the technology standard BATNEEC rather than any measure of environmental capital. Sustainability management systems and standards are at a very early stage of development, are not yet subject to external certification or verification, nor referenced to any measure of environmental or social capital. TNS provides a sustainability framework, but no time scale for compliance. All these initiatives are voluntary and unlikely to be taken up by companies unless there is perceived competitive advantage in doing so.

Despite the continued and increasing pressures on businesses to formulate and implement environmental and sustainability policies, and evidence that some are enthusiastically engaging with these agendas, there is a school of thought which holds that, left to itself, the business sector will always do much too little, too late. Korten (1995) explains the unsustainable nature of current global development in terms of the power of multinational corporations (MNCs) to extract value for their shareholders unsustainably from the environment and from the poor. This power is both economic and political and, for as long as it is retained by MNCs, Korten claims, depletion of environmental and social capital by business cannot be reversed. Only if national governments and citizen coalitions challenge and take back control can sustainable development be achieved.

These arguments imply that the measures described in this chapter can never make enough of a difference and that government intervention will be needed to bring about the necessary reforms to curb the size, influence and rapaciousness of big business.

Chapter 6 now examines how policy, including environmental policy, is made and implemented by governments at the national level.

Further reading

An introductory text to corporate environmental management is Hitchcock and Blair (2000). Howes *et al.* (1997) takes a detailed look at the drivers for improved performance. Sheldon (1997) is a series of essays covering a wide range of practical and theoretical issues surrounding the development of the ISO 14000 series. Elkington J. (2001a) describes the emerging sustainability agenda and how companies might respond. For a US perspective see Erickson (1999) and for information on corporate environmental management in the developing world Utting (2002). The argument that for business to become sustainable we need fundamental change in the control of the business sector is set out in Korten (1995) and Klein (1999).

6 ▸ Environmental policy making in government

- The policy-making cycle at national government level
- Theories of power and the role of interest groups
- The rational-comprehensive, incrementalist and mixed scanning accounts of decision making
- Regulatory, economic and persuasive policy instruments
- The role of evaluation in policy making
- The prospects for policies of sustainability

Environmental policy in government

As the focus shifts towards larger scale policy making, the complexity both of the policy making organisations, and of their spheres of influence, makes the study of policy making more complicated, but also fascinating and rewarding. At governmental level policies are not necessarily adopted in order to achieve a set of discrete objectives that are well understood. Within political systems, where vested interests operate with varying degrees of influence and success, the stated aims of a policy may often mask its real purposes. There is rarely only one point of decision and therefore an easily identifiable 'policy maker'. Even less rare is one coherent set of policies, structured so that one never contradicts another. This chapter explores these complexities and introduces some of the models that have been used to analyse them.

Environmental policy at the national level

As with corporate policy making (Chapter 5), the policy making process can be broadly conceptualised as a cycle (Figure 6.1) although the reality is unlikely to be quite as tidy as the figure suggests (Jenkins 1978). The five main components of the figure (policy environment, inputs, government, outputs and outcomes) are introduced and discussed in turn through the rest of this chapter. Note that the terms 'resources' and 'environment' used in Figure 6.1 have different meanings here from those that were used in previous chapters of this book.

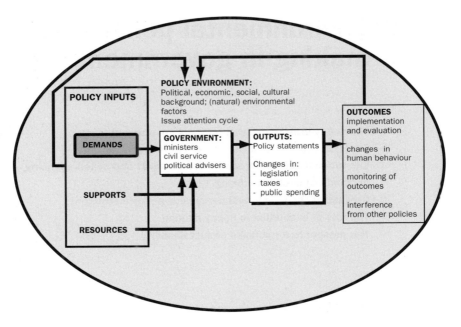

Figure 6.1 *The policy-making process at government level*

Policy environment

The policy environment is the context in which the policy system is operating at any one time. It includes the political complexion of the government of the day and prevailing public political ideologies. Economic circumstances also shape the policy environment. Social and cultural factors, including prevailing attitudes and values (Chapter 2), also contribute.

'Environment' in the sense more often used in these pages is also important here. Information about the state of the natural environment, and evidence of emerging and established environmental problems, form part of the policy environment. The way this information is presented, mediated and interpreted is of course crucial.

Downs (1972) has suggested that public and media attention follows a predictable cycle for most issues, including environmental problems (Figure 6.2). After a period in which the problem exists but is not given much attention (stage 1) the problem emerges into the public sphere and enters the political agenda of the day. This emergence generates alarm and any proposed solutions are greeted enthusiastically (stage 2). Inevitably, it then emerges that the problem is more intractable than

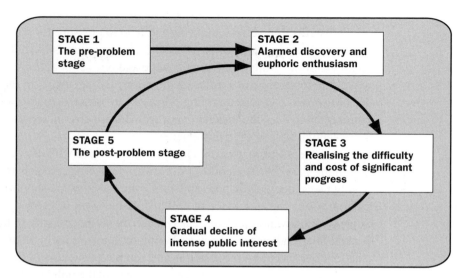

Figure 6.2 *Downs's issue attention cycle*
Source: Downs (1972)

originally thought: the proposed solutions may turn out to be expensive, or difficult to implement, or ineffective, or all three (stage 3). The resulting stalemate leads to a tailing off of enthusiasm amongst the public (stage 4) – although those with a particular and long-standing interest in the issue may remain as committed as ever. The issue then slips into relative obscurity, where it is unlikely to receive much attention (stage 5) until new evidence of its severity or effects emerges and the cycle is re-started.

For environmental issues, the model generally seems to provide an explanation of the waxing and waning of issues within the policy environment. There are many examples, from deforestation of the Amazon and global warming to acid rain, where public opinion has reacted in more or less the manner described. However, Jordan and O'Riordan (2000) suggest that the cycle has speeded up, and the stages have become less distinct, in the thirty years since Downs first proposed it.

Inputs

Easton (1965) proposed that in modern democracies policy is made in response to the pressures, opportunities and constraints provided by a combination of three classes of inputs:

- *Demands*. These arise within the policy environment. Perceived problems affecting various interests will lead to pressure for policies to be formulated to address the problem. This pressure is likely to be complex. Some groups may want change and one set of actions; others will argue for change in a different direction; and yet others to retain the *status quo*. The structures and processes by which demands are represented to decision makers in government are considered in much more detail a little later in this chapter.
- *Supports*. Democratic governments will only survive with, as a minimum, the passive support of the governed. Activities such as paying taxes, obeying the law, and voting are the baseline supports without which the legitimacy of policy decisions will be eroded. Supports are given to the political system and not necessarily to the personalities and parties in power. Voting for a losing party in an election, but accepting the result, is an example of support of this kind. Losing such support from even a minority can make policies difficult or impossible to sustain, as events in the autumn of 2000 showed in Britain and in France. Protests by lorry drivers and farmers, against what were seen as excessive transport fuel duties, closed roads and oil refineries in both countries, leading to widespread economic disruption.
- *Resources*. The most important resource for any policy is often the money required to implement it, and the availability of this will constrain policy formulation. However, other resources may also be significant: human (the availability of workers with the necessary skills to implement a policy); information (for example scientific information as discussed in Chapter 4); and technical (the availability of the right technology). The availability of natural resources and other environmental services will sometimes be a consideration and sometimes the major focus of a policy.

Power, access and representation

Figure 6.3 is an expansion of the 'demands' area of Figure 6.1 and the arrows show the routes by which representations are made. The figure is based upon the central components of the UK government, but similar diagrams can be drawn for other democratic nation states. The formal constitutional route for representation is shown in the upper section of Figure 6.3. Citizens provide support for the system by voting and can make demands by lobbying their elected representative. Political parties provide an organising framework for this process by publishing election manifestos with alternative sets of policies. Parties choose candidates for

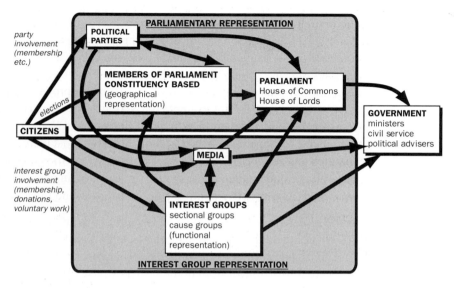

Figure 6.3 *Interest groups and representation*

parliamentary elections and the party with the most seats in the House of Commons forms the government. Between elections citizens can raise issues with the parties or with MPs, who can in turn make representations to ministers or civil servants. These may or may not result in changes to policies.

However, between elections the policy debate on most political issues revolves around the organised interest groups that have a pivotal role in representing particular points of view to the government, both directly and via the media. There is fierce competition between groups and some are much more successful than others in achieving their objectives. There are two types of interest group, sectional groups and cause groups.

● *Sectional groups* represent the interests (usually economic) of a particular class of people. Trade unions, professional bodies and employers' organisations are obvious examples. However, some sectional groups are less easily identifiable, because making representations within the political system is only a small part of their activities. Most companies, for example, will make political representations when necessary, for example if legislation relevant to the company's activities is being drafted. The main UK motoring organisations exist to provide their members with benefits such as breakdown recovery and insurance, but often are to be found lobbying government on issues such as road building and the taxation of motoring.

- *Cause groups* promote issues that are not directly related to the interests of their members. Environmental groups such as Greenpeace and Friends of the Earth (Box 6.1), groups campaigning against poverty (domestic or international) or animal protection organisations are all examples of cause groups. Like sectional groups, some cause groups direct most of their energy not to campaigning but to providing services, either to their members (for example the National Trust, a charity which preserves the built and natural heritage of England and Wales) or to other clients (for example Oxfam). Some, however, exist mainly to campaign and services to members are limited to activities which will support this campaigning, either directly by encouraging and enabling members to take an active political role, or through fund raising.

The distinction between sectional and cause groups is not always clear-cut. An environmental group campaigning against a motorway extension, landfill site or quarry might be assumed to be a cause group – but if the members are mostly drawn from home owners whose properties would lose value if the development went ahead the true interest is sectional. NIMBY (Not In My Back Yard) is the acronym used to describe this specific class of groups.

The power to influence decision making is closely related to the ability to make effective representations to policy makers (Figure 6.3). Groups can lobby opinion formers within political parties; individual Members of Parliament directly and through their membership base; and Members of the House of Commons and House of Lords collectively, for example Select and Standing Committees of the House of Commons. They will also attempt to lobby the ministers, civil servants and policy advisers who make up the inner policy-making circles of the government. The groups themselves often interact, seeking to forge alliances and affect the strategies and tactics of each other.

Groups use the media by manufacturing 'news items' in the form of press releases and conferences, reports, leaked government documents (if they can get them), or stunts such as demonstrations, occupations etc. Press, broadcasting and other media organisations choose which stories to run, how much prominence to give them and how to present each side's arguments and personalities, and so have a crucial role in the process overall. The media cannot be assumed to be passive and neutral transmitters of interest group communications. They often have their own political agenda, shaped to appeal to a particular target audience or by the business interests or political affiliations of their proprietors.

Meaningful discussions with representatives of the government are available to relatively few groups. Powerful lobby groups that possess such access are called insider groups – some of these may be so powerful that they hardly need to use the less effective routes of the media and Parliament. The rewards of such an influential position within the political system are great, but they come with some costs. Insider groups must seek consensus, not confrontation, with government if they are to retain their status. The leaderships of these groups often have to spend as much time selling compromises to their membership as negotiating these with the government. A group will be admitted to insider status only if it meets some of the following conditions:

- *Authority*. A track record of authority and fair dealing which reassures the government that it will not abuse insider status.
- *Compatibility of objectives*. The interests promoted by the group must be of importance to the government either economically (e.g. large companies; trade unions); or electorally (e.g. fashionable cause groups).
- *Usefulness*. The group must offer government something in return for insider status. Specialist information, professional expertise, and moderation of its members' demands are all positive attributes that can help a group into government circles.
- *Sanctions*. Sometimes groups become insiders because governments fear what they may do if not allowed consultation and influence. Examples here include powerful media interests, which could produce adverse publicity for the government if not appeased, and trade unions in essential industries with the power to strike. (Jones 1998: 170)

Most environmental issues involve decision making by more than one government department and some groups may be insiders in one ministry but excluded from others. Connelly and Smith (1999: 76) give the example of the Council for the Protection of Rural England (CPRE) and its campaign against major road-building projects. Accepted into consultations by the then Department of the Environment, the group was excluded from the Treasury and had only limited access to the then Department of Transport. For outsider groups with no special access at all an important strategic consideration is whether to work towards insider status or whether the freedom that comes from being an outsider group outweighs the potential benefits of improved access to decision makers. Box 6.1 examines the strategies of two UK environmental groups. Friends of the Earth and Greenpeace have chosen different paths in this respect but each has grown in effectiveness none the less.

Box 6.1

Friends of the Earth and Greenpeace

During the 1970s and early 1980s the newly established Friends of the Earth (FoE) and Greenpeace were unambiguously outsider groups, relying on eye-catching media stunts, public rallies and letter-writing campaigns to get their message across. The 1971 Schweppes bottle dump, when FoE supporters 'returned' several thousand non-returnable bottles to the London headquarters of Cadbury Schweppes, launched the group in the United Kingdom. One of Greenpeace's first actions in the United Kingdom was to use inflatable craft to harass the dumping of waste from Britain's nuclear power programme into the Atlantic. Both these events generated arresting visual images and hence much press and television coverage.

Over the years, both groups have evolved from being purely outsiders to a situation where they participate, to some extent, in insider lobbying. FoE has consistently chosen to cultivate the necessary respectability and expertise to engage with government, whilst still pursuing legal direct action on some issues. Greenpeace, by contrast, has chosen to continue to emphasise direct action in its campaigns. Although it will engage with government on occasion, its preferred route is pressure from the outside.

A snapshot from the anti-nuclear campaigns of both groups in 1983 illustrates well this difference in approach. FoE spent the whole of that year engaged in a public inquiry into the application to build Sizewell B nuclear power station in Suffolk. This huge endeavour (the inquiry lasted twenty-seven months and FoE's costs were £250,000) put a huge strain on the organisation's human, logistical and financial resources. The outcome was a decision to build the station – seemingly a defeat for FoE. By contrast, Greenpeace was campaigning to prevent the dumping of UK nuclear waste at sea. Publicity stunts involving the Greenpeace boat *Cedar Lea* in the Bristol Channel were complemented by negotiations with the transport trade unions, which were eventually persuaded to instruct their members not to transport the waste by train or by boat. No nuclear waste from the United Kingdom has been dumped at sea since that year.

Hinkley Point C was to have been the next nuclear power station built after Sizewell B. The inquiry started in 1988 and ran for just over a year. FoE were again represented, but this time Greenpeace also played a significant role in the inquiry, preparing its own evidence whilst also funding a local advice centre for objectors. Legislation to privatise the electricity industry was going through Parliament at exactly that time and the contrast in styles of the groups is illustrated by the uses each made of its participation in the inquiry. Whilst FoE sought double value from its inquiry evidence by presenting it to the House of Commons Environment Committee, Greenpeace's strategy was extra-parliamentary. Its aid for local objectors was aimed at getting as many individuals and groups to object as possible, thus delivering a powerful message to the government and to potential investors in the electricity industry.

Both the Sizewell B and Hinkley C inquiries were, on the face of it, battles lost by the anti-nuclear movement, but in a war which, for the time being, they seem to have won. Although permission was given to build both power stations, only Sizewell B was constructed and at the time of writing there are still no firm plans to construct any

Box 6.1 continued

further nuclear stations in the United Kingdom. Hinkley C and its planned sister stations were abandoned, in large part owing to the very high costs of nuclear waste management. The different campaigning strategies of FoE and Greenpeace over the previous decade had complemented each other. FoE's tenacious engagement at public inquiries had delayed the programme overall by ensuring that, for each proposed station, the gap between application and consent was a matter of years, not months. Greenpeace's closing off of the sea dumping route forced the industry to seek expensive land-based disposal methods, a process which is still not resolved for high-level waste.

References: Lamb (1996); Pearce (1991); Roberts (1991).

Discussion points

1 Does participation in a public inquiry indicate that a group has achieved insider status?

2 How could participation in the Sizewell and Hinkley public inquiries help FoE to meet one or more of the conditions for insider status?

3 If the nuclear power policy sector is characterised by corporatism, is insider status for FoE a realistic possibility?

Models of interest group representation

The relatively simple concept of insider and outsider groups, with different forms and levels of access to decision makers, can be elaborated into models of government–interest group interaction. These aim to demonstrate the relationships between groups and government and the processes by which representation and other communications occur. The models are used to characterise particular policy sectors, defined broadly (for example, the environmental policy sector) or more narrowly (for example the nature conservation policy sector or even the small reptile preservation policy sector).

The pluralist model describes systems where no groups have strong insider status, the political system is open, and power is dispersed. Whilst there is not equality of access, no group is excluded from having its say. Government acts as an arbiter between the different groups and takes the major role in implementing its policy decisions once these are made. Pluralism more often describes policy sectors where the issues are ethical rather than economic, although some analysts have argued that it characterises most group–government interaction in states such as the United States and United Kingdom (Dahl 1961; Jordan and Richardson

1987). Abortion laws, divorce, the age of consent, homosexuality and penal policies are the usual examples cited of issues where pluralism is the 'best fit' model of the policy sector.

The corporatist model describes systems in which insider groups are in close co-operation with government, not just making policy but involved in its implementation also. In return for this influence, they accept compromises on their demands and constraints on their behaviour, including the behaviour of their rank and file. Policy is mostly made behind closed doors, with groups bargaining with each other and with government until a consensus is reached and can be presented to the legislature and the citizenry at large. Power is tightly concentrated in this model: outsider groups are largely excluded from influence and make little headway.

Policy sectors where significant economic interests are at stake tend to resemble the corporatist model (Schmitter 1979; Cawson and Saunders 1981; Wolfe 1977). Examples here are policy sectors concerned with energy, the chemical industry, other large manufacturers and engineering firms, and agriculture – note the key role many of these have in the use of environmental capital. However, competing models have been developed to account for the concentration of power within political systems.

Élite theory proposes that small numbers of individuals control most decisions (for example ministers, senior civil servants, those with great scientific or technical expertise and those at the top of party hierarchies) and that groups will be successful to the extent that their interests coincide with those of the élite. *Marxism* is a class-based analysis, which suggests that the government is the agent of the capitalist class, i.e. those who own and control the means of production. Political decisions will always further the interests of this class either directly or indirectly (Miliband 1969).

The utility of these models is to allow analysts to characterise, categorise and compare the processes of policy making in different sectors. Even at the level of the policy sector, real life will rarely fit neatly into one model or another, but the extent to which policy sectors are characterised by inequalities in access and power is a key question for students of the policy process, as well as for the actors within it.

The limitations of the pluralist/corporatist models when applied to large scale policy making have led some analysts to study individual policy sectors in order to identify and characterise the policy networks which operate within them. This approach allows a subtler characterisation of the relationship between the groups themselves and groups and government, or parts of government such as ministries and agencies, than the rather crude insider–outsider approach. 'Policy communities' is the term used to describe close-knit policy networks within which groups are relatively powerful. 'Issue networks' describes the relationships of groups within more open policy sectors, where groups' potential for influencing policy outcomes is correspondingly less. Table 6.1 sets out the characteristics of these two types of policy network.

Power without representation

Many studies of the political process are based upon observations of interest group activity. However, some potential actors in the policy process may be able to exert power without doing anything at all.

A classic study of environmental policy making by Crenson (1971) demonstrated that powerful vested interests could control the political agenda through the anticipated reactions of policy makers. Crenson carried out a comparative analysis of the development of air pollution regulation in two US cities, Gary and East Chicago. Both are steel-making towns, but whereas in East Chicago there were several steel plants owned by various corporations, in Gary the industry was almost wholly owned by US Steel. East Chicago introduced air pollution regulations in 1957, Gary only in 1962 in response to federal legislation. Studying government and media records, Crenson found no evidence that US Steel had used its economic power to delay changes in the law. However, Crenson suggests that the authorities anticipated that, if regulated, US Steel would close its plant and move elsewhere, and that this pre-empted any debate within Gary on this issue. The anticipated reactions of the more fragmented industry in East Chicago were less intimidating to policy makers and hence regulation was easier to introduce.

Determining the distribution of power within this policy sector by studying the actions of groups would give a fundamentally incomplete account of the relationships. Power would appear to be distributed along pluralist lines, when in fact the system is élite. The players who were in fact the most powerful were relatively inactive, because they had no need to lobby. According to Crenson, policy making in this situation is characterised by non-decision making, meaning failure to raise issues in the first place.

Lukes (1974) has developed the analysis of non-decision making and suggested that there are three 'faces of power'. First is the overt activity of interest groups within the political arena; second is the control of the

Table 6.1 Types of policy networks: characteristics of policy communities and issue networks

Dimension	Policy community	Issue network
Membership		
Number of participants	Very limited number, some groups consciously excluded from network	Large numbers of groups; open structure
Type of interest	Economic and/or professional interests dominate	Any issue, including those where economic and other (e.g. environmental) issues overlap
Integration		
Frequency of interaction	Frequent, high-quality, interaction of all groups on all matters related to policy issues	Contacts fluctuate in frequency and intensity
Continuity	Membership, values and outcomes persistent over time	Membership and structure fluctuate significantly
Consensus	All participants share basic values and accept the legitimacy of the outcome	A measure of agreement exists, but conflict is ever present
Resources		
Distribution of resources (within network)	All participants have resources (information; legitimacy; control over implementation); these are used to bargain for influence	Some participants may have resources, but they are limited, and basic relationship is consultative
Power		
Within participating organisations	Hierarchical; leaders can deliver members	Varied and variable distribution and capacity to regulate members
Between participating organisations	There is a balance of power among members. Although one group may dominate, the network as a whole is stronger than its individual members would be on their own	Unequal powers, reflecting unequal resources and unequal access. Winners and losers exist: the winners at the expense of the losers, so that there is no overall gain in power or benefit to the network as a whole

Source: adapted from Marsh and Rhodes (1992: 251)

political agenda via anticipated reactions, so that the grievances of outsiders are not addressed. The third is the manipulation of citizens' preferences so that they fail to recognise grievances in the first place and hence cannot express them. Conditioned by socialisation, education and the media, people do not necessarily recognise and act upon their own 'true' interests. This echoes Max-Neef's analysis of 'pseudo-satisfiers' in Chapter 2 and is open to the same criticism of paternalism. Both Max-Neef's and Lukes's analyses rest on two assumptions. One is that people's objective needs and interests are not necessarily the same as the conditioned and subjective needs and interests that individuals express. The other is that academics or policy makers can identify these objective needs and interests – that is, that only experts know what ordinary people really want.

Political system

After the representation stage, the next phase in the policy-making process is the decision-making process within government (Figure 6.1). In every modern democracy, this will take place within a complex structure, comprising elected politicians, civil servants, insider group representatives and advisers from the ruling party or parties. Politicians are ultimately responsible for policy decisions, but detailed work on policy proposals is carried out by civil servants. Insider groups may assist in this task by providing information and expertise. Another important source of advice and influence is the teams of political advisers to ministers and secretaries of state, although such political staff may not be formally part of the policy-making process.

The methodology by which decisions about policy are made has consequences for the type of policy outputs that result. Three models of decision-making methods are introduced below:

- rational-comprehensive decision making;
- incrementalism;
- mixed scanning.

These models have two potential uses. One is to describe how decision making within organisations (including government) actually takes place in order better to understand and analyse this process. As with the models of interest group representation introduced above, the utility of the model in these terms depends how well it is able to illuminate actual decision making processes. No model would ever be able fully to capture the complexities of real-life organisations and the human beings who work within them: but by identifying key features of organisational processes useful insights can be identified. The second function is normative, or prescriptive. Analysts can evaluate and recommend some decision-making methods as being more likely to result in sound decisions than others.

All three techniques require predictions to be made about the probable effects of different policies by evaluating the policy at the formulation stage. Methods of policy evaluation are introduced later in this chapter.

Rational-comprehensive decision making

The rational-comprehensive decision making model was developed by Simon (1945) as a means of characterising organisational decision making, which includes decision making within governments. The model proposes that decision makers first formulate a set of goals, then consider all possible courses of action by which the goals might be met. They should then select the one that is most likely to achieve the desired goals. Problems with applying the model to real life include:

- *The rationality of the goals.* In governmental decision making the goals adopted are strongly influenced by a political process involving the jostling of interest groups with varying power and access, rather than the objective analysis suggested by the word 'rationality'.
- *The complexity of the rational-comprehensive method.* Although the method is easily described, carrying out an analysis using the method would be so time-consuming and expensive that no policies would ever get made. Lack of information and the difficulty of predicting the effects of policy changes are additional difficulties (Ham and Hill 1993).

Incrementalist decision making

Incrementalism is an alternative model, which proposes that policies develop in small, incremental steps, rather than in fundamental leaps as suggested by the rational-comprehensive approach. The starting point for the incrementalist approach is the existing situation, rather than an idealised future goal. A limited number of potential policy adjustments are compared with each other and the set which is considered most likely to secure the support of stakeholders (because improvements in the existing situation are predicted) is adopted. The situation is monitored and further changes are made if deemed necessary as time passes. In this way, the model makes clear the role of values in the decision-making process. The model suggests that policy making is based on negotiation and consensus with affected interests, rather than objective rationality.

Lindblom (in Braybrooke and Lindblom 1963) advocated the incrementalist model as both descriptive and a prescriptive ideal. The model was claimed to be much better suited to the limited resources and means of analysis available to real people in real organisations than the rational comprehensive method. Furthermore, the use of negotiation (termed 'mutual partisan adjustment') would result in decisions that were fair between parties and therefore acceptable. Other analysts, for example Dror (1964) however, have criticised the method as inherently conservative. Adjusting the *status quo* step by step, it is claimed, cannot produce radical solutions to deep rooted problems. Every journey starts with one step, and a multitude of small incremental changes can add up to significant change over time. However, adopting the incrementalist method does limit the options available to decision makers by ruling out more radical options. Given the radical nature of policy changes that will be necessary to achieve sustainable development, this is an important drawback to this method in the context of environmental policy making.

Decision making by mixed scanning

Mixed scanning (Etzioni 1967) aims to overcome this limitation by assessing the nature of the policy problem before deciding on a method. It suggests that fundamental and long-term decisions are best undertaken by a broad-brush review of the predicted consequences of a set of policy alternatives. This is not the detailed analysis required by the rational-comprehensive model but bounded by considerations of cost, time and the availability of information. However, radical as well as incremental policy changes can be considered in this review. The result is a strategic decision about the general direction which policy should take: incremental decisions, based on detailed analysis of more limited changes, then fine-tune the direction of the policy until it is judged time for another fundamental review.

In Box 6.2 various models of policy making are illustrated by the story of the Non-fossil Fuel Obligation (NFFO), a policy measure introduced by the British government to, inter alia, increase the proportion of electricity generated by renewable energy.

Box 6.2

The non-fossil fuel obligation

In 1988 two cherished policies of the UK government were in opposition. Long-standing plans to build a series of pressurised water reactor (PWR) nuclear power stations were about to come to fruition. The first of these, Sizewell B, had just completed its long inquiry stage and construction was getting under way. However, the government also wanted to privatise the electricity supply industry, which meant that investment in future PWRs must come from private, rather than public, sources, and it was highly unlikely that this could be achieved. Box 8.1 explains why the characteristics of nuclear power make it an unattractive investment in a free market situation.

The coal industry strikes of 1974 and 1984 had convinced the Prime Minister, Margaret Thatcher, that expanded nuclear capacity was needed to reduce the United Kingdom's dependence on coal as a fuel for electricity generation. To achieve this in a privatised industry it would be necessary to offer investors in PWRs a guarantee that they would be able to sell electricity from the stations in the future whatever the prevailing economic conditions.

The non-fossil fuel obligation (NFFO) was initially conceived to fulfil this need. The NFFO required the companies supplying electricity to purchase from generators a certain quantity of electricity from non-fossil sources. Companies were to be compensated for the extra expenditure they incurred through a levy on electricity generated from coal, oil and gas. A leaked draft of the policy had proposed a 'nuclear obligation'. Protests that renewables could be as effective as nuclear in providing diversity and security of supply meant that in the final draft renewables were included and the obligation was renamed. As renewable capacity in England and Wales at that time was negligible, it was anticipated that the main beneficiary would anyway be the nuclear industry.

Initially the government planned to increase the NFFO year on year in order to force the construction of nuclear power stations to meet the increased requirement. But problems quickly emerged. The European Commission, concerned that the NFFO gave nuclear power an unfair advantage under competition laws, imposed an eight-year time limit on the NFFO. This meant it could be of little use even to Sizewell B (eventually commissioned in 1994), let alone any PWR not yet under construction. In late 1989 the PWR programme was effectively abandoned by the government and all nuclear power stations were withdrawn from the privatisation proposals.

The NFFO, therefore, demonstrably failed in its stated intention, to facilitate the construction of nuclear power stations in a privatised electricity supply industry. However, it has been very successful in securing private investment in renewable energy schemes, such that the proportion of electricity in England and Wales generated by renewables has risen from virtually nothing in 1990 to 1,200 MW in 2003 (DTI 2003). In addition, the competitive bidding arrangements for NFFO contracts have resulted in sharp falls in the price of renewable electricity, so that some technologies, such as onshore wind power, are now competitive with fossil electricity in some circumstances.

Box 6.2 continued

One important factor in this success was the EC decision that NFFO support for renewables could be continued after the nuclear industry ceased to be eligible in 1998. NFFO contracts were awarded in five rounds between 1990 (NFFO 1) and 1998 (NFFO 5). Civil servants in the Department of Trade and Industry (DTI) used tranches within the NFFO mechanism to create protected markets within markets. Thus wind energy projects bidded against each other for NFFO contracts, the projects with the lowest price succeeding. Waste-to-energy schemes similarly competed, as did biomass schemes. As experimental renewable technologies approached viability, NFFO tranches were created to encourage them, for example the tranche allocated to the anaerobic digestion of farm wastes in NFFO 4.

Further restructuring of the electricity market in 2000 abolished the NFFO, although existing contracts run on for a maximum of fifteen years. However, the emergence of global warming as an international policy issue (Box 2.2) has led to several European countries adopting policy measures to encourage the replacement of fossil-fuelled electricity generation with renewable sources and the NFFO model has been under active consideration by some countries in continental Europe.

References: Mitchell (2000); Roberts *et al.* (1991).

Discussion points

1   From the NFFO story, can you identify examples of (a) *rational-comprehensive*, (b) *incrementalist*, (c) *mixed scanning* methods of policy making?

2   Which were successful and which unsuccessful?

Policy outputs

To be implemented successfully environmental policies need to be able to change people's behaviour – to change the pattern of activities that are depleting environmental capital. The outputs of the policy system are policy instruments – actions which governments can use to effect such changes. People will change their behaviour only if forced to by laws and rules; if they perceive an economic advantage in doing so; or if they are persuaded that they ought to take action of their own accord. To illustrate this, it is useful to think back to the 'tragedy of the commons' problem in Chapter 2. Three ways were suggested for the commoners to take collective action to avoid the 'inevitable'. They could establish a council or government and pass laws to regulate the behaviour of individuals; they could decide to tax those who over-used common resources in order to discourage unsustainable use; or they could use social encouragement and coercion to discourage antisocial behaviour.

These three classes of policy instrument are in fact all that there is in the policy maker's armoury, no matter what scale of organisation is being considered. Policy instruments are therefore classified into laws and regulation (sticks), economic instruments (carrots) and persuasion, ranging from straightforward provision of information through education through to propaganda. In most policy contexts a combination of the two or all three instruments is used. For example, to discourage excessive speed by motorists the government imposes speed limits (regulation), fines those who break them and are caught (economic instruments), and runs press and television advertising campaigns to persuade drivers to comply.

Laws and regulations

Regulation through law is either the direct result of legislation passed by elected legislatures, such as the British Parliament or American Congress (primary legislation); or results from rules (secondary legislation) drafted by ministers or officials acting under powers granted by primary legislation. As a policy instrument, well drafted legislation can have many advantages and is widely used. If legislation can be straightforward; is introduced quickly in response to emerging problems; sets out clear and fair rules that are equitable between stakeholders; and is enforceable, it is likely to be the most appropriate policy instrument to use.

In many contexts, however, the legislative approach on its own will not be ideal. If the change in behaviour required is absolute (for example, householders are forbidden to dispose of their waste by fly tipping) legislation is appropriate. However, if it is a matter of degree (householders should recycle more) legislation is rather a blunt instrument. Central government may use legislation to force local government into action on this issue, maybe by requiring the production of recycling plans by each local authority. However, the policy instruments in these plans which will directly impact on householders are unlikely to be regulatory in nature: persuasive or economic measures will have more chance of success.

A key issue is enforcement: laws which are expensive, difficult or impossible to enforce are best avoided, not least because they may bring the whole legal system into disrepect. This applies even if the change in behaviour is absolute rather than incremental. To pass a regulation requiring householders to recycle certain materials (and therefore prohibiting, say, aluminium cans from dustbins) would require enforcement officers to check the contents of bins and apply sanctions in response to non-compliance. Whilst not impossible to implement, this would be time-consuming and might alienate public opinion.

Similar considerations apply to the regulation of pollution emissions by companies and corporations. The traditional approach to pollution control has been to prohibit certain polluting processes unless prior licensing and consent have been obtained by the company from a government regulatory agency. If the authorising legislation is strict enough, stringent control by regulators is then possible through the licence agreement. As well as specifying limits on actual emissions, licences can be used to mandate the type of plant used and the way it is operated; the design and implementation of emergency procedures; staff numbers, qualifications and training; and even, under the European Union's Integrated Pollution Prevention Control Directive, the adoption by the company of an environmental management system, including energy efficiency and waste minimisation plans.

There are, however, inherent limitations in regulation as a policy instrument for pollution control. Firms are likely to respond to licence conditions in a way that complies – but only just. Regulation gives no incentive to invest in pollution reduction beyond the legal limit. Pollution control regimes that insist on 'best available technology' often enforce the requirement at the time that technology is upgraded. The extra capital expense involved in state of the art technology can make extending the life of older, polluting, equipment more attractive, especially as 'state of the art' can sometimes mean technologically immature and unproven.

Regulation is also often inflexible and economically inefficient. Imagine a company with ten factories each producing the same amount (say ten units per year) of the same pollutant, X (Figure 6.4). Total annual pollution is therefore 100 units. If pollution licensing is introduced each will be given a licence requiring a cut in emissions of the same amount, say 20 per cent. This will require capital investment by the company at each of the ten plants, probably in 'end of pipe' technology to remove pollutants from the emissions immediately before their release. The result of this investment is ten sites producing eight units each, giving reduced emissions of eighty units per year.

However, the same sum of money invested in completely new plant at four of the sites might result in 90 per cent emissions reduction for each of these four but no reduction at the other six plants. Total pollution is then sixty-four units per year (six producing ten units each and four plants each producing one unit). This is a more economically efficient outcome, as a greater reduction has been achieved for the same capital outlay. Increasingly economic approaches to pollution control are being adopted

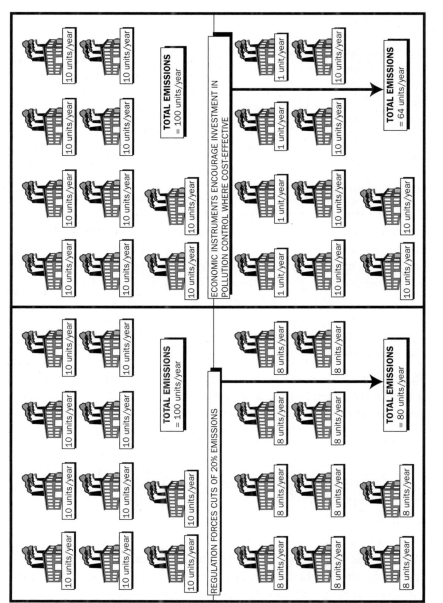

Figure 6.4 Regulatory and economic systems of pollution control

by governments keen to achieve environmental objectives in the most economically efficient way, for example the 1990 US Clean Air Act (US EPA 1993).

Economic instruments

Economic policy instruments can overcome some of the disadvantages of regulation by providing a more flexible and potentially more efficient means to encourage polluters to reduce their polluting behaviour. Examples of economic instruments include pollution charges, product charges, tradable permit systems and government subsidies – the full range and its theoretical basis are discussed in more detail in Chapter 8. They are not an alternative to regulation but a more sophisticated form of it. Unless voluntarily agreed upon (and this is rare for reasons discussed in the next section), economic instruments are imposed by legislation to levy a pollution tax, product charge or subsidy and provide for its collection and enforcement (or allocation and distribution in the case of subsidies).

In the ten factory example (Figure 6.4), if a charge for each unit of pollution produced is imposed, managers will have an economic incentive to reduce emissions, provided the cost of doing so is less than the charge. Each site manager will then calculate the most cost-effective response, depending on the individual circumstances of the plant. Pollution charging provides a continuing incentive for firms to reduce emissions as, for example, technological improvements bring down the cost of pollution control. The lower the emissions, the lower the cost to the firm – unlike licences based on maximum emissions targets.

Economic instruments can be problematic if the activity being taxed meets basic needs or makes a significant contribution to the quality of life of economically disadvantaged groups. Economic instruments that raise the price of water, food or energy for heating or transport, for example, will have a proportionally greater effect on lower income groups than on more affluent consumers. If this increases poverty or leads to basic needs not being met the sustainable development principle of equity is violated. In these cases, subsidies may be appropriate to counteract the effects of the charge. For example, if the cost of motoring is raised beyond the reach of low-income groups, subsidies to public transport can provide an alternative and affordable mode of transport. Raising domestic energy prices through taxation might be justified in the context of subsidies to improve the domestic energy efficiency of vulnerable groups.

Persuasion and voluntary action

For a government seeking to implement a particular environmental policy, getting individuals or companies to change their behaviour without the need for regulation, taxes or subsidies is an attractive proposition. As well as being cheap (no expensive monitoring or enforcement required), it also offers a way of initiating environmental action through consensus. This has the potential to increase the acceptability of the particular action being taken, and possibly environmental action in general, and maybe even the popularity of the government by association.

Voluntary action works best where the behaviour change is relatively easy to make and costs little or no money. Recycling of domestic refuse is a good example of an issue where many people may make significant changes to their behaviour with just a little persuasion. If the action required is time-consuming or costs money, however, people will be reluctant to join in. This is not just because of the inherent difficulty of doing so. In classic tragedy-of-the-commons style, the question 'Why should I, if others can't be bothered?' will act as an added deterrent to action. Firms will have a similar attitude, which is why economic instruments are rarely adopted voluntarily. Even if the majority of companies in a particular industry agree to impose, say, a product charge to fund the environmentally safe management of their product at the end of its useful life, they are likely to be undercut by 'free riders' – the minority of manufacturers who choose not to join in and can therefore offer the goods at lower prices.

All persuasion rests ultimately on information and there are various styles of information provision that may be appropriate in particular circumstances.

- *Passive information provision*. Information materials containing neutral information are made available but only to those who enquire. This course of action has the advantage of cheapness but is unlikely to have much impact on behaviour. It might be used where the government has reluctantly provided a subsidy in response to a pressure group campaign but is happy for there to be a low uptake.
- *Active information provision*. Advertisements and/or mail shots are used to bring the information to the attention of its target audience in a promotional campaign.
- *Active persuasion*. This begins to move beyond the provision of neutral information – arguments are used to engage emotions and make those not wishing to comply feel guilty.

- *Propaganda*. Appropriate where there are important social reasons why certain types of behaviour must stop, for example drink-driving. In this example the main deterrent is regulation – the propaganda acts to alter the social consensus so that drink-driving becomes less acceptable and therefore less likely. The behaviour change being sought is disapproval of those who might drink and drive so that social consensus reinforces the legislation. Taken to extremes, this results in coercion and potential violation of human rights.

Achieving sustainable levels of births in developing countries is a particularly difficult policy problem. The behaviour to be changed is intimate, private and often governed by strong cultural and religious beliefs. In Box 6.3 the effectiveness of mix of policy instruments which have been used to implement population control policies in China is examined.

Policy outcomes

Whereas policy outputs are government actions, policy outcomes are the actual effects (impacts) of these actions and therefore of the policy in the real world. These complete the policy cycle that was introduced in Figure 6.1 by changing the policy environment and contributing to changes in the nature of demands, supports and resources. These changes, however, are not necessarily the ones desired and planned by policy makers. Policies can fail at the implementation stage for all sorts of reasons – resistance by civil servants or other bureaucrats; resistance by citizens; or simply because the policy instruments chosen were inappropriate. There are many examples where implementation of one policy prevents another from being successful. In the early 1990s privatisation of the UK electricity industry stalled the nuclear power building programme, both of which were cherished government policies (see Box 6.2).

Even if the desired change in behaviour happens once a policy is in place, it might not be entirely due to the policy itself, but to the unintended effects of another. The shift from coal to gas as the main fuel for electricity generation in the United Kingdom during the 1990s was due more to the privatisation of the power stations (Roberts *et al*. 1991) than to policies to limit acid emissions, although these played their part (see Box 7.1).

Box 6.3

Policy instruments for population control

In 1979 the Chinese government adopted a one-child-per-family policy. Although its success was not uniform across the country, the policy did achieve relative success in achieving its main target. This was to limit the rapidly rising Chinese population to 1.2 billion by the year 2000. The actual end-of-century figure was 1.26 billion and by that time the annual population growth rate had fallen to less than 1 per cent (World Bank 2002).

The policy was implemented through the layers of national, provincial and local governments and down to Communist Party cadres in urban and rural communities. No couple could bear more than one child unless they had been given specific permission to have a second. There were two types of gestation in China – authorised and unauthorised. The stated policy was based on voluntarism – that is, that it would be achieved through individuals exercising their own reproductive choices with guidance from the state. In fact there was coercion to secure compliance, and to 'rectify' unauthorised pregnancy through abortion. Each of the three types of policy instrument (regulatory, economic, persuasive) was used to implement the policy.

Regulation

The 1980 Marriage Law, which set a minimum age for marriage of twenty-two for men and twenty for women, was the prime national implementing legislation. Pregnancy in women younger than twenty was automatically 'unauthorised'. There were other supporting laws, one of which made it a criminal offence to remove an intra-uterine contraceptive device (IUD) from a woman without official permission.

Some flexibility was handed down to provincial governments on implementation, provided overall regional targets were met. Provincial family planning regulations specified the circumstances in which couples might be granted permission to have more than one child, for example if the first child was disabled.

Economic instruments

Couples complying with the policy might be granted preferential access to land, food, housing, health care and educational opportunities for their offspring. In urban areas compliance could bring salary increments. In rural areas agricultural producers' compliance was linked with the ability to trade. Farmers could sell their produce only by signing two linked contracts – one agreeing to the terms of trade and the other agreeing not to start an unauthorised pregnancy. The penalty for violation was loss of income from crops. Other economic disincentives included fines, loss of free social services and demotion at work. Economic incentives and disincentives also applied to cadres and higher-level officials if the birth target of the community for which they were responsible was met or exceeded.

Box 6.3 continued

Persuasion

The persuasive tactics used to enforce the one-child policy ranged from information provision to extreme coercion. Contraceptives were distributed to women of childbearing age and education on how to use them effectively was widely available through workplaces and the community. Meetings were organised to persuade people of the need for birth control and of the link between small families, wealth generation and a better quality of life.

Married women were intrusively monitored by birth control workers, in some cases to the extent of checking that they were menstruating each month. If an unauthorised pregnancy was detected cadres might then exert relentless pressure to have an abortion. Similar pressure was put on women with two or more children to undergo sterilisation.

Conclusion

The dilemma facing the Chinese government is clear: should the reproductive rights of the present be generation be respected even if it leads to disaster for future generations? Malthusian logic (see Box 2.2) is particularly persuasive in the Chinese case as the area of arable land per capita is only one-third of the global average. Stabilising population growth in these circumstances is urgent. In seeking a balance between individual and collective interests the decision has been weighted towards the latter, resulting in the serious infringements of personal freedom noted above. The justification of the policy lies in its effectiveness in reducing the rate of population growth, thus protecting future generations (Wolf 1986: 115). However, some commentators (e.g. Hartmann 1995: 170) claim that coercion need not be a necessary component of an effective birth control programme: persuasion linked with improvements in women's health care, education and status, could do more to improve the quality of life of men and women whilst respecting human rights (see Box 1.3).

Reference: Hardee-Cleaveland and Banister (1988).

Discussion point

1   Identify the regulatory, economic and persuasive features which characterise the implementation of family planning and sexual health polices in one or more countries with which you are familiar.

Policy evaluation

Evaluation of policies once implemented is an important component of the overall policy process. This is especially the case if policies have been expensive (either to the government or to other parties) or if they have entailed limitations of personal liberty. Policy evaluation is not merely retrospective, however: it provides information, which can be used to improve implementation by adapting policies themselves or policy outputs. For this reason, formative evaluation (at the policy formulation or reformulation stage) is as important as summative evaluation (undertaken when a particular policy has run its course). For any particular policy, evaluative research can be undertaken to determine:

1 To what extent were the actions of implementing officials and target groups consistent with the objectives and procedures outlined in that policy decision?
2 To what extent were the objectives attained over time, i.e. to what extent were impacts consistent with the objectives?
3 What were the principal factors affecting policy outputs and impacts, both those relevant to the official policy as well as other politically significant ones?
4 How was the policy reformulated over time on the basis of experience?

(Sabatier 1986:22)

Methods of evaluation must be appropriate to the circumstances. These will range from quantitative techniques – for example large statistical surveys or economic techniques such as cost–benefit analysis (see Chapter 8) – to qualitative research (for example on public attitudes to drink-driving following a television advertisement campaign). Baseline data will always be needed, so planning for evaluation needs to be thought through at the time the policy is formulated, not tagged on as an afterthought.

Making policies for sustainability

Policies for sustainable development have to incorporate as objectives the protection and enhancement of environmental, social and economic capital. What does the above analysis of the policy process tell us about the chances of developing policies that comply with sustainability principles? The message from the issue-attention cycle is a depressing one, suggesting that public opinion will acclimatise to accept dire environmental and other problems – provided that the consequences are reasonably distant in terms of time or space.

Analysis of interest group representation suggests that the representation of non-economic interests, such as environmental protection and sustainable development, will be usually at a structural disadvantage compared with economic interests. Cause groups, such as environmental pressure groups, are more likely to make headway when policy sectors are structured on pluralist lines, because corporatist structures will usually mean outsider status for these groups. However, as Table 6.1 shows, issue networks are characterised by pluralism and open access but little power for groups within them.

This is not to say that progress towards sustainable development is impossible in the face of vested economic interests. The last three decades of the twentieth century saw huge improvements in the environmental standards of industry in the United States and European Union, especially with regard to waste and pollution. This progress was in no small way due to effective lobbying and campaigning by environmental groups. This has resulted in changes in government policies, tighter regulation of industry and, as was seen in Chapter 5, the development of voluntary action by businesses in their enlightened self-interest. These developments have happened incrementally, demonstrating that significant changes can occur in many small stages. Technocentric analysts are likely to be content with this approach; ecocentrics would prefer faster and more radical changes.

Achieving the equitable distribution of the fruits of environmental capital seemingly remains an intractable problem at both national and international levels. The problems of poverty and social exclusion in developed and developing countries are not solved by increased environmental protection, nor even necessarily by the production of more material goods. The Brundtland Commission (WCED 1987: 8) noted the need for 'effective citizen participation in decision making and . . . greater democracy in international decision making' to help to ensure the needs of the poor were met.

The short-termism inherent in electoral politics impedes the development of policies which impose costs on the present generation in order to benefit future generations. This has led some analysts to advocate authoritarian government by an 'enlightened' élite as the only way of imposing policies for sustainability (e.g. Ophuls 1977; Heilbroner 1977). Others argue for decentralisation and subsidiarity, reasoning that the concomitant empowerment of individuals and communities would lead to the cultural changes and enhanced sense of personal responsibility which are necessary for individual behaviour to change (e.g. Bookchin 1974; Illich 1973).

Democratic systems do not provide for the systematic representation of the interests of future generations. Most of the potential beneficiaries of sustainable development policies have not yet been born and hence cannot found and support interest groups to press for policy changes. It therefore falls to enlightened cause groups in the here-and-now to act on their behalf. Persuading decision makers, and the citizens who elect them, that it can be necessary to defer immediate benefits for the sake of future generations is a difficult task.

The development of economic instruments to implement environmental policies in the 1980s and 1990s has potentially made the implementation of environmental policies more efficient in economic terms. However, significant issues of equity can arise if the result is to price basic commodities beyond the reach of the poor. It seems clear that the scale of change required to achieve sustainable development will only be accomplished if the hearts and minds of most sections of society are won over to the necessity of change. This suggests that the key policy instruments will be neither regulatory nor economic (although these will have significant parts to play) but persuasive.

This chapter has reviewed the policy process at the national level. Increasingly, however, the environmental policy of nation states is becoming an important component of their foreign policies. Chapter 7 examines these trends and how environmental policy is brokered at the international level.

Further reading

The classic texts on the various stages of policy making in government are referenced in the text above, for example Easton (1965), Downs (1972), Dahl (1961), Schmitter (1979), Crenson (1971), Lukes (1974), Braybrooke and Lindblom (1963). Up-to-date summaries, commentaries and critiques of these works, set in the context of environmental policy, are found in texts such as Connelly and Smith (1999) and Hawke (2002). Gray (1995) provides a review of a range United Kingdom government environmental policies. For an overview of environmental pressure group activity in the UK see Rawcliffe (1998). Chapter 2 of Radcliffe (2002) reviews the debate between the advocates of authoritarian and decentralised political structures.

International environmental policy making

- Features of international politics relevant to sustainable development
- The issues of globalisation, debt and trade in the context of global development
- The global processes of negotiation on environment and development

National sovereignty and international law

The 'tragedy-of-the-commons' model, introduced in Chapter 2, can be used to illustrate problems relating to open access resources at any scale. Although the model is set at the level of conflict between individual households in a hamlet, it can exemplify conflict of interest in the management of common resources between communities at a urban scale; regions at a national scale; and nation states at the international or even global level. However, the model has particular resonance when applied either to individuals or to nations.

Men, women and children have rights, enshrined in the Universal Declaration of Human Rights. In the absence of government, which is an assumption built into the tragedy model, humans have autonomy to act as they please without taking into account the effects of their actions on others. This anarchic state is more difficult to imagine when the model is scaled up to, say, the level of towns and cities, as such developments are unlikely to arise in the absence of government. But, used to describe the management of the global environment by the different countries of the world, the model fits very well indeed. This is because nation states are sovereign – just like a medieval dairy farmer, within their own borders they have the right to do as they please even if this damages environmental capital upon which other nations depend, such as the atmosphere and the sea.

International law, of course, does exist, but one of its fundamental tenets is the principle of national sovereignty and this is much more widely respected in practice than is the doctrine of universal human rights. There is no single international body with enforcement powers. For environmental policy making on the global scale this presents some formidable problems. To circumvent the 'tragedy', the three broad types of policy instrument are available: legal, economic and voluntary action. But, in contrast to intra-national policy, laws and taxes cannot be imposed on sovereign nations without their prior consent to this through treaty agreements. This means policies tend to be diluted to a level acceptable to the least enthusiastic nation. This problem is exacerbated by the relative lack of sanctions available to punish free-riders – it is possible for countries to gain significant competitive advantage through continuing to pollute whilst other nations assume the costs of cleaning up domestic industries. Although international law allows tort (damages) claims by one state against another these are difficult to prove in commons-type cases. For example, damage from trans-boundary air pollution may be difficult to link conclusively with one particular geographical source, both spatially and temporally. Box 7.1 illustrates these issues in relation to acid pollution in Europe.

All is not anarchy on the international stage, however, because nations recognise the advantages of acting collectively to avoid or at least ameliorate environmental problems. Some states voluntarily pool some aspects of their sovereignty with others, the European Union (EU) being a good example of this. Sharing sovereignty is usually done in the interests of trade, but common environmental policies are often a result of such unions. Minimum standards for all are needed to ensure fair competition. This is one reason why this chapter is as much concerned with trade as it is with environmental management – the other reason being the role of trade in the quest for sustainable economic development. In Box 7.2 the implications of unregulated trade liberalisation between developed and developing nations are examined for the particular case of fishing.

Even when states remain autonomous, negotiations on issues such as trans-boundary air pollution or fisheries management lead to treaties which, once ratified by signatory countries, become binding in international law. Such negotiations will share many of the characteristics of national policy making that were outlined in Chapter 6. Decision makers at the international level operate in a policy environment which generates demands, supports and resources which are processed to produce policy outputs and outcomes (Figure 6.1). The scale of decision making is larger, the interest groups more diverse and negotiations conducted in several different languages, but at the most basic level the system is the same. Just as at the national level, international policy is a dynamic process, more (or less) open to interest group participation, but inevitably more accessible to some groups than to others.

Box 7.1

Controlling sulphur emissions in Europe

The link between sulphur emissions elsewhere in continental Europe and the acidification of lakes in Scandinavia was first suspected in the 1960s. Subsequent research demonstrated that atmospheric acid emissions could travel hundreds or thousands of kilometres before falling as acid rain, snow or dry deposition. As evidence of widespread ecological damage in Europe emerged it became clear that a co-ordinated international response to trans-boundary acid pollution was required.

Policy on this issue has developed in parallel within two forums: the United Nations Economic Commission for Europe (UNECE) (open to all European countries plus the United States and Canada because some acid pollution is transatlantic) and the European Union (formerly the European Community (EC)).

United Nations Economic Commission for Europe

The 1979 UNECE Convention on Long-range Trans-boundary Air Pollution acts as a framework for monitoring emissions, research into their effects, and co-ordinated international strategies for emissions reduction. Emissions reductions are enforced through protocols to the convention, which are legally binding on only nations that ratify them. For sulphur emissions the key protocols are:

- *1985 Helsinki Protocol on the Reduction of Sulphur Emissions or their Trans-boundary Fluxes*. Only twenty-one countries are parties to this protocol, which required 30 per cent cuts in sulphur emissions (or trans-boundary fluxes) by 1993. Negotiations were complicated by scientific uncertainty about the mechanisms linking emissions, deposition and ecological damage. The United Kingdom declined to ratify the protocol, claiming that more scientific information was necessary before incurring the substantial economic costs involved in emissions reduction. In contrast, the then Federal Republic of Germany based its approach on the precautionary principle (Chapter 4).
- *The 1994 Oslo Protocol on Further Reduction of Sulphur Emissions*. By the 1990s scientific models had been developed which allowed predictions to be made of the effect that emission reductions from any particular source would have on ecological quality elsewhere. The 1994 protocol therefore takes an effects-based approach. Priority was given to emission reduction from the most harmful sources, with the eventual aim of protecting even the most vulnerable areas (for example peaty uplands) from damage by acid rain. Inevitably, some countries had to accept higher emission reduction obligations than others. Despite this, most significant polluters, including the United Kingdom, ratified the protocol.
- *The 1999 Gothenburg Protocol to abate Acidification, Eutrophication and Ground-level Ozone*. This protocol sets national emissions ceilings for sulphur emissions (as well as nitrogen oxides, volatile organics and ammonia). Effects on human health, as well as environmental quality, are taken into consideration. The basis of the

continued

Box 7.1 continued

negotiated ceilings is both scientific and economic. Nations whose emissions are the most harmful, and those which would be able to reduce them relatively cheaply, have had to promise the largest cuts.

European Union

At the same time as the Helsinki Protocol was under negotiation, the European Community was developing a directive on sulphur emissions from power stations and other large fossil fuel plant, the Large Combustion Plant Directive (LCPD). Using 1980 as a baseline, the original proposal was for a 60 per cent reduction in sulphur emissions from such plant by 1995 across all member states. The United Kingdom and other member states that would find it difficult to comply, such as Italy, Greece, Eire, Spain and Portugal, strenuously resisted this. After tortuous negotiations (Boehmer-Christiansen and Skea 1991: 234–46) the final version of the directive set different proportional targets for different member states. For the United Kingdom these were 20 per cent by 1993, 40 per cent by 1998 and 60 per cent by 2003.

In response to improvements in scientific understanding, and in parallel with the same shift in the UNECE negotiations, by the 1990s effects-based policies were preferred. The Acidification Strategy, proposed in 1997, sought further reductions of key pollutants, including sulphur, in order to reduce the area within the European Union at risk from acidification from 6.5 per cent on the basis of existing commitments to 3.3 per cent by 2010. Implementation of the strategy will be by:

- *Sulphur Content of Certain Liquid Fuels Directive*: setting maximum levels for the sulphur content of heavy fuel oil and gas oil.
- *Amendment to the EC Large Combustion Plant Directive*: to further reduce emissions of sulphur, nitrogen oxides and particles from new large combustion plants.
- *National Emissions Ceiling Directive*. This gives legal force within the European Union to the UNECE 2010 targets for sulphur dioxide, nitrogen oxides, ammonia and volatile organic compounds.

References: Boehmer-Christiansen and Skea (1991); DEFRA (2000), chapter 3; UNECE (2002).

Discussion point

1 To what extent does this case study show that sovereignty is an inevitable barrier to solving commons-type problems between nation states?

Box 7.2

Senegal: fish, trade and sustainability

As the coastal waters of developed nations become more regulated to prevent overfishing and depletion of fish stocks, their fishing industries have sought new waters in which to operate. Many developing countries have welcomed foreign fleets, foreseeing that the increased national income would assist with debt repayments and that the related economic activity would provide jobs and income for their citizens. Research by the United Nations Environment Programme (UNEP), however, has found that where there is insufficient regulation of foreign fishing fleets, severe environmental, economic and social detriment to the host country can result.

The West African state of Senegal has become particularly dependent on this source of export income since the 1980s. Encouraged by trade liberalisation measures guaranteeing access to European markets, fish exports grew rapidly. By the late 1990s two-thirds of the country's overseas earnings came from fish. Keen to boost its exports, the Senegalese government encouraged the trade through agreements with foreign fleets. Export subsidies of up to 25 per cent further encouraged the trade. When Senegal was forced to devalue its currency the exports increased further as the fish was even better value in European markets.

Senegalese waters are now overfished and some stocks, especially those from coastal deep waters that are popular with European consumers, are in short supply. Some may be in danger of complete collapse. Fish is an important source of protein for local people, who are now in competition with the foreign fleets for a diminishing resource. Food shortages in local markets are predicted. Local wildlife dependent on fish – for example whales, dolphins and seals – are additional competitors and at risk of following the fish stocks into population collapse.

The problem has been made worse by the failure of Senegal to fully capitalise on its fish resource by insisting that catches were processed locally in up-to-date plant. Most export stocks are of raw or frozen, rather than processed, fish. Inefficient handling and the out-of-date plant used for the catches that are processed locally mean high wastage rates, increasing pressure on stocks.

The UNEP research recommends policies that would allow Senegal and similar countries to continue to trade fish internationally but with safeguards to protect national economies and the marine environment. The sustainable management of fish stocks is essential. To achieve this UNEP recommends:

- increased access charges for foreign fleets, which would have the double benefit of fully compensating Senegal for the costs of fishing whilst reducing the demand for what would become a more expensive resource;
- access agreements that can be suspended should scientific evidence emerge that stocks are endangered;
- the imposition of quotas for each species, to preserve those most at risk;
- diversification of exports to other parts of Africa and to Asia, where consumer tastes are different from those in Europe, thus evening out the pressure on different fish species;

continued

Box 7.2 continued

- economic incentives to encourage investment in modern processing plant so that more value is added to the fish resource before it is exported;
- economic incentives, such as lower fuel prices, for small fishing boats serving the local markets.

Reference: UNEP (2002).

Discussion points

Developing countries such as Senegal have little political power when negotiating with other governments, international organisations such as the WTO, and multinational corporations.

1   How does this relative powerlessness affect the likelihood of the UNEP recommendations being implemented?

2   What would be the effect if Senegal were the only developing country to introduce these measures?

3   How could Senegal improve its negotiating position with the rest of the world?

A world of two halves

The paradox of development in the twentieth century was that, as economic, technological and cultural change swept the world, seemingly reducing the diversity of different continents and peoples, the very same developments were widening the gap between the haves and the have-nots. In tandem, political changes have occurred. The accelerating reliance on market forces and consumer driven solutions in the last two decades of the twentieth century has empowered some individuals but reduces the ability of organisations and governments to plan for the collective good.

The very concept of sustainable development was born from the need to remedy global inequalities. To do this will need understanding of the linked issues of globalisation, the indebtedness of poorer nations, international trade and environmental quality at the regional and global scales.

Globalisation

Globalisation emerged in the last decades of the twentieth century as an over-arching concept to explain rapid economic, political, cultural, social and technological changes that were underway throughout the world (Hoogvelt 1997; Giddens 1999). The fundamental drivers of globalisation are transport and telecommunications technologies, which have reduced, and are continuing to reduce further, the barrier that distance represents. This has facilitated the creation of a global economic system, replacing the more regionalised models that pertained as little as thirty years ago. This is not to claim that international trade in itself is new: what has changed, as a result of globalisation, is the volume and nature of that trade.

This is due to the growth of multinational and transnational companies (or corporations or enterprises). Multinationals (MNCs) operate relatively autonomous businesses in several different countries, whereas transnationals (TNCs) have highly integrated operations working across national frontiers. In the past, goods branded with the logo of a US company would almost certainly have been manufactured in the United States, perhaps using imported commodities such as oil or steel. These days, branding gives no clue to the origins of products. Transnationals contract with a multiplicity of manufacturers all over the world – some to produce components, some to assemble them into products and others to package and dispatch to wholesalers the finished goods. Even very basic manufactured goods, such as clothing, are likely to have passed through several different countries before reaching the shops.

In economic terms, each region is both in competition with, and interdependent upon, the whole world. Countries with a tradition of high wages and good social security benefits find their exports undercut by low wage economies, in neighbouring countries or in other continents. Worker militancy in one place can lead to the transfer of investment to somewhere more promising. The tendency of market-based economic systems to produce inequality is explained in Chapter 8 and this has been the outcome of the growth of globalised capitalism. Although wealth generation has increased rapidly for the world as a whole, this growth in prosperity has been unequally shared.

In the second half of the twentieth century the world economy grew fivefold. Despite the concomitant growth in world population, mean *per capita* income grew 260 per cent in the same period. However, the ratio of income between the richest and poorest 20 per cent of the world population doubled from 30:1 to more than 80:1 between 1960 and 1995. In 1993 more than 1300 million people had less than US$1 a day to live on (UNEP 2000b). Although most of them lived in Asia, the Pacific, sub-Saharan Africa and Latin America, poverty in the industrialised world, especially but not exclusively the former Soviet bloc, affected significant minorities of the population of most developed countries. Inequalities in wealth correlate reasonably well with inequalities in resource consumption and waste production. Figure 7.1 shows estimates of variations between regions' average environmental footprint.

Debt and development

One important feature of this growing inequity was the indebtedness of many developing countries. The roots of the debt crisis which came to a head during the 1980s and 1990s lay in the rapid rise in oil prices which followed the 1973 Arab–Israeli war and the 1978 Iranian revolution (Corbridge 1993). This, coupled with the prolonged US trade deficit during this period, resulted in large offshore holdings of the US currency, mostly in European banks. The economies of the developed world were in recession following the oil price hike and investment was depressed. Large companies were tending to raise money by issuing bonds rather than borrowing from banks. So the banks sought other potential debtors to lend to – and many developing countries proved easily persuadable.

Often, loans were used to fund consumption of imported goods (including sometimes armaments) rather than on infrastructure that would yield a return. In those cases where loans had been used to construct capital assets the resulting projects were not always successful, even in economic terms. When measured against sustainability criteria, they became even more questionable. Large dams, to take just one example, whilst producing electricity for cities or industrial sites such as aluminium smelters, often resulted in the displacement and impoverishment of peasant farmers from large areas of flooded land (Adams 2001: 233–40). In addition, the environmental impact of these projects was often large and adverse. Dams disrupt the hydrology of river basins, altering the pattern of drought and of flooding downstream. Valuable habitats and eco-systems can be lost and the decomposition of submerged vegetation leads to large-scale emissions of methane, a powerful greenhouse case (Box 1.2).

The debt crisis that resulted from this rapid lending by western banks to developing countries was dramatically manifested in 1982 when Mexico defaulted on its loan repayments, although earlier signs of emerging difficulties could be found in the debt renegotiation by several countries in earlier years. Rising US interest rates in the early 1980s had a double effect on indebted countries, raising both the amount of local currency needed to repay the dollar debt and the annual interest charge. As a succession of countries ran into similar difficulties, rescue packages were

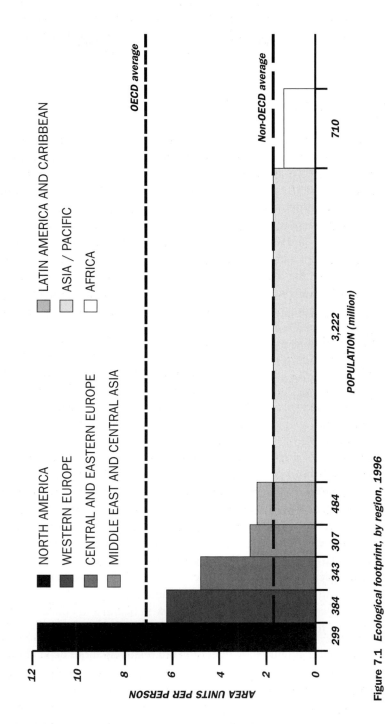

Figure 7.1 Ecological footprint, by region, 1996

Source: UNFPA (2001), by courtesy of the UN Population Fund, New York

developed to prevent wide spread default. The extent of lending by this stage was so great that, had developing nations simply refused to pay, the stability of the banking system world-wide would have been threatened.

The response of the lending countries to the crisis was to do all that was necessary to prevent the collapse of their banks and the severe economic consequences this would entail. As debtor nations ran into difficulties servicing their original debts, rescheduling arrangements were put into place, usually brokered by the International Monetary Fund (IMF) (see Table 7.1).

As well as stretching the originally agreed term of the loans well into the future, to reduce annual capital repayments, further loans were imposed, in order to improve short-term cash flow. This gave both lenders and debtors some breathing space but was not, of course, seen as the long-term solution. Householders who fall into debt must reduce their consumption and increase their income to avoid bankruptcy. Using the same logic, the Structural Adjustment Programmes (SAP) imposed by the IMF as a condition of rescheduling included requirements to decrease public spending; reduce imports of goods and services; and, wherever possible, increase exports of goods and services. There the analogy ends, however. Householders can escape their debts by going bankrupt and, after a period of time, have their debts written off and make a new start. There is no such provision for indebted nations – debts, once incurred, run on, for generations if necessary, until they are either forgiven or repaid.

SAP policies have had effects that run entirely contrary to the three key principles of sustainable development. Rather than producing equity, inequality has been exacerbated, both between nations and within nations. The imposition of SAP regimes led to countries falling into a spiral of increasing indebtedness, with large proportions of GNP earmarked simply for repayments to foreign banks. Decreased welfare spending impacted most severely on the urban and rural poor, as health and education spending was cut back. Future generations will be adversely affected – economically, as those born after the original debts were incurred and spent will be liable for repayments under the rescheduled arrangements; socially as the reduced investment now in health and education will have impacts for generations to come; and environmentally.

Developing countries by definition are relatively more dependent on commodities, rather than manufactured goods, for export earnings. The pressure to increase exports accelerated a ruthless and short-term approach to commodity extraction, resulting often in environmental damage as well as long term resource depletion – for example see Box

Table 7.1 *International bodies and agreements relevant to international development, trade and environment negotiations*

DFID	Department for International Development (UK)	http://www.dfid.gov.uk/
EU	European Union	http://www.europa.eu.int/
FCCC	Framework Convention on Climate Change	http://unfccc.int/
GEF	Global Environment Facility	http://www.gefweb.org/
IMF	International Monetary Fund	http://www.imf.org/
IPCC	Intergovernmental Panel on Climate Change	http://www.ipcc.ch/
G7	Group of Seven (Canada, France, Germany, Italy, Japan, United Kingdom and United States)	http://g8.market2000.ca/
G8	Group of Eight (G7 plus Russia)	http://g8.market2000.ca/
G77	Group of 77 (representing less developed nations)	http://www.g77.org/
GATS	General Agreement on Trade in Services	http://www.wto.org/
GATT	General Agreement on Tariffs and Trade	http://www.wto.org/
NAFTA	North American Free Trade Organisation	http://ffas.usda.gov/info/factsheets/nafta.html
OECD	Organisation for Economic Co-operation and Development	http://www.oecd.org/
UN	United Nations	http://www.un.org/
UNCED	United Nations Conference on Environment and Development	
UNECE	United Nations Economic Commission for Europe	http://www.unece.org/
UNEP	United Nations Environment Programme	http://www.unep.org/
WTO	World Trade Organisation	http://www.wto.org/

7.2. Further examples include the unsustainable logging of rainforests in South America and Indonesia (Pierce-Colfer and Resosudarmo 2001); pesticide pollution from the horticulture of flowers and luxury vegetables in East Africa (Maharaj and Dorren 1995); and oil exploitation in Nigeria leading to pollution of soils and water courses (Ikein 1990). Each of these is an environmental problem with serious consequences for future, as well as present, generations. Low commodity prices in global markets compound the problems as countries have to produce greater volumes of commodities each year to maintain their export earnings. These lower prices are in part caused by the increased global competition between exporting countries stimulated by SAP.

Figure 7.2 demonstrates the deterioration in the debt position of all developing regions save for Latin America and the Caribbean between 1980 and 1995. Pressure on western governments to resolve the situation, in part from development NGOs, led to a series of international initiatives to write off the debt of the most indebted countries. Initially, small-scale initiatives were introduced to convert loans to grants. These were followed in 1996 by the Heavily Indebted Poor Countries (HIPC) programme. This aimed to ensure that each poor country was relieved of its unaffordable debt burden, provided it could demonstrate sound economic management. In 1999 the HIPC programme was broadened by lowering the threshold of debt deemed to be unaffordable and including an eligibility requirement that a national poverty reduction strategy must be in place before countries qualify for relief.

By July 2001 twenty-three countries had met the HIPC conditions but there were signs that even the 1999 debt threshold might prove too demanding for some of these, risking a further downward spiral of unaffordability. Other nations were excluded from the programme through being unable to meet the conditions in terms of economic management or poverty reduction. Development groups represented by the Jubilee Movement International for Economic and Social Justice continued to call for a total write off of all poor country debt. However, even if this was to be achieved, it would not necessarily prevent further debts from being incurred in the future. Sustainable wealth creation in the developing world is needed if future indebtedness is to be avoided. It is therefore necessary to examine the rules that govern trade between nations and the extent to which they assist or hinder sustainable economic growth.

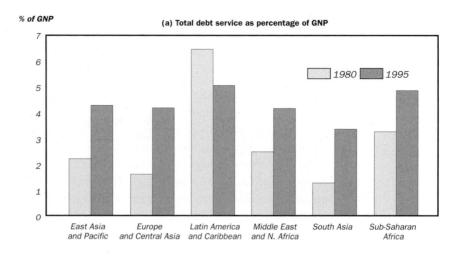

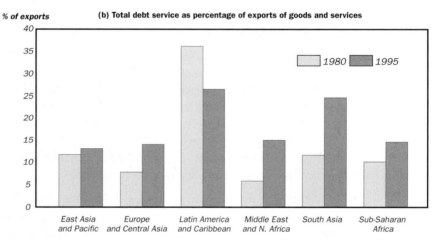

Figure 7.2 *Debt in the context of GNP and exports, 1980 and 1995. (a) Total debt service as a percentage of GNP. (b) Total debt service as a percentage of exports of goods and services*

Source: Department for International Development (1997), by courtesy of HMSO, Norwich, Crown copyright

Trade

International trade is crucial to the developing world. According to classical economic theory, healthy competition and liberalised international trade will promote economic growth and international trade – and the solution to poverty is wealth creation. During times of global recession poorer countries suffer disproportionately as the demand for their commodities falls along with the volume of exports and unit price.

Chapter 8 explains in more detail the theory of free markets and their role in terms of economic efficiency and wealth creation. However, it also argues the incompatibility of a truly free market approach and the three key sustainability criteria – equity, futurity and valuing the environment.

The potential for trade to create wealth and diminish poverty is obvious – but in practice international trade can have the opposite effect, as Box 7.2 demonstrates. It all depends on the rules – how they are made, what they require and how they are implemented. Rules are necessary to stop unfair practices, for example high import tariffs, that act as a barrier and give unfair advantage to domestic industries. This is clearly something which countries at equivalent stages in terms of their development should avoid – if country A imposes high tariffs on one set of goods, countries B, C and D may do the same. The stage is then set for an escalating and damaging trade war in which the advantages of trade are lost to each participant.

However, it can be argued that the more vulnerable economies of developing nations should be allowed to protect nascent or even established domestic industries at least in the short- to medium-term. After all, the industrial development phase in Europe and North America took place behind protectionist tariff barriers. Government subsidies to agriculture, industry or the service sector are another potential source of unfairness and similar arguments apply – subsidies for exported goods are inherently unfair but exceptions can be justified to encourage the development of particular sectors in poorer countries.

The body with responsibility for regulating international trade is the World Trade Organisation (WTO). Negotiations started at the end of World War II resulted in the General Agreement on Tariffs and Trade (GATT). During the period 1986–94 the so-called Uruguay Round of GATT negotiations brought about significant liberalisation of the agreement and, on 1 January 1995, the establishment of the WTO. The organisation's functions are:

- administering the WTO trade agreements which have superseded GATT;
- acting as a forum for trade negotiations;
- settling trade disputes;
- reviewing national trade policies;
- assisting developing countries in trade policy issues through technical assistance and training programmes;
- co-operating with other international organisations (WTO 1999).

The benefits claimed for the system administered by the WTO are that it promotes trade and efficient wealth creation and that it allows trade disputes between nations to be settled according to agreed rules and procedures without recourse to threats of war (WTO 1999). However, the operation of GATT and the WTO has been much criticised by those who claim that the rules are unfair to the poor and damaging to the environment (e.g. WDM 2001). Insisting on the progressive removal of tariff barriers exposes the developed economies to healthy competition but may prove fatal to new and vulnerable industries in the developing countries. WTO rules on environmental and consumer safety standards specify maximum, rather than minimum, standards and thus have the effect of pulling all nations down to the same floor. Placing all producers on to the same global playing field makes no allowance for the power imbalances that exist between small producers (peasant farmers, or small enterprises in developed or developing nations) and multinational companies. This power stems from the economic might of the MNCs but its nature is both economic and political. Vested interests in Western countries, such as farmers or steel producers, are also more powerful than their developing world counterparts.

During November 2001 a new round of WTO negotiations, called unsurprisingly the Millennium Round, started and is likely to take three or more years to complete. The outcomes from the talks will be fundamental to the prospects for the developing nations.

Agricultural subsidies in the European Union are likely to be reduced or phased out entirely as part of the negotiations. This has been pressed for by developing nations, as well as the United States, Canada and Australia. The effect on European consumers would be that locally produced food would probably become more expensive and imported food relatively more attractive. This could allow greater penetration of European markets by developing nations – but only in competition with North American and other large scale agricultural producers.

An issue that is particularly sensitive for the United States is tariff barriers. Under US law, import duties can be imposed if the government believes that goods are being sold at below the cost of production. This is called anti-dumping legislation and has been contentious in the case of several commodities including steel, with countries such as Japan, Brazil as well as the European Union complaining that, in fact, the United States is protecting its own steel making industry unfairly. American trade unions will resist any change to anti-dumping laws on the grounds that US jobs may be lost if they are scrapped. There is a wider issue though of national sovereignty. Unlike the European nations which have agreed to pool sovereignty within the overarching structures of the European Union, the United States jealously guards its freedom from outside interference and will resist attempts by the world community, through the WTO, to enforce change to its internal laws.

Investment in services is another topic raising concerns about sovereignty. During the 1990s the OECD proposed a Multilateral Agreement on Investment (MAI). This would have given multinational companies the right to set up businesses in any signatory country and to claim compensation from that country if these businesses were adversely affected by local laws, for example on labour rights or environmental protection. An inevitable consequence of such an agreement would be the levelling down of environmental and social standards to the minimum, at least as they applied to multinational companies. The proposal was abandoned in the face of campaigning pressure from development groups and concerns of OECD members about their own sovereignty. For example, France withdrew support for the MAI, worried about the impact on the French language and culture of uncontrollable foreign investment. Although multinational investment in production is now on the Millennium round agenda, similar concerns, from both developed and developing nations, are likely to make agreement very difficult.

However, the GATT agreement covered only commodities and manufactured products: the Millennium Round will entail discussion of the General Agreement on Trade in Services (GATS). This seeks to liberalise investment in services such as telecommunications, banking, tourism and insurance. It may extend to sectors which some countries choose to run on a publicly funded basis such as water supply, health, and education. This has led to concerns that current subsidised provision in the developing world could be privatised and bought by Western companies, putting it beyond the reach of the poorest (WDM 2001). The counter argument is that competition will reduce prices and benefit consumers (WTO 2001), but this is little consolation in the case of services which previously were free at the point of demand.

Intellectual property rights are also addressed, especially in the context of pharmaceutical patents. Again this is an issue where the interests of the developed and the developing countries are opposed. Multinational companies are keen to prevent software piracy and other copyright violations and these are common in some parts of the developing world. Some new drugs can be virtually unaffordable – not because the cost of manufacture is high, but because the medicine is still under patent to recoup the research and development costs of the pharmaceutical company which developed it. Developing countries are seeking blanket exemptions from these restrictions in some circumstances, such as a major public health emergency.

Lastly, the Millennium Round will have to reconcile the big differences between nations on the issue of trade and the environment. Sustainable development, optimal use of the world's resources and environmental protection were among the founding principles of the WTO. However, governments seeking to restrict imports on the grounds that their production has been environmentally damaging have been impeded by WTO rules. For example, the US ban on Mexican tuna, which was caught using nets which also trapped and killed dolphins, was overturned by the WTO. The rules state that such bans must be justified by scientific evidence – ethical concerns, such as animal welfare, or consumer concern, as in the case of genetically modified organisms, are not sufficient. As was seen in Chapter 4, the nature of science and the scientific method mean that it is often difficult, when environmental problems arise, to 'prove' a link between factors causing the problem and the problem itself.

The European Union in particular will be arguing for a much higher priority to be given to environmental protection in the rules which emerge from the Millennium Round, including the adoption of the precautionary principle (Chapter 4) so that the balance of decisions made in advance of scientific certainty will swing towards environmental protection rather than free trade. In this, they are likely to be opposed by many developing nations who foresee discrimination against their products due to their lack of access to up to date 'green' technology and to the funds needed to install it. Suspicion that the motivation for such rules is less environmental and more about gaining unfair advantage for developed countries with spare money to spend on environmental protection is natural. As Maslow's theory (Chapter 2) suggests, environmental concerns may be the prerogative of the affluent. Even if less developed countries accept the principles of sustainable development, the dire poverty of their people forces upon them shorter term solutions to achieve wealth creation.

One factor influencing the outcome of the Millennium Round will be public opinion in the developed world, where voters and consumers are increasingly aware of the links between debt, trade and poverty. The 1990s saw a sharp rise in the availability and popularity of 'fair-traded' goods in mainstream retail outlets in Europe. Brokered and certificated by NGOs, fair-trade schemes provide long-term contracts to small producers of commodities such as coffee and chocolate, provided certain standards on wages, child labour and environmental protection are met. Just as the Jubilee 2000 campaign put pressure on western governments on the issue of debt relief, so it is likely that issues of inequalities in trade relations will assume a rising political profile during the ongoing world trade talks.

Environment

In parallel with the WTO negotiations on trade, there is a range of international bodies whose role is to broker agreements on environmental issues at either the global or the regional level. They are outlined in Table 7.1. Foremost amongst them is the United Nations (UN) and its subsidiary bodies, for example the UN Environment Programme (UNEP). The 1992 UN Conference on Environment and Development (UNCED) was a watershed event, marking as it did the widespread acceptance by governments worldwide of the concept of sustainable development as a means of reconciling the development imperative with the need for environmental protection, both now and in the long term future. That progress towards global sustainable development has subsequently proved limited and difficult does not diminish the significance of the event.

Also known as the Rio Conference or Earth Summit, UNCED brought together delegations from 172 nations, the majority of which were led by the head of state or head of the government. There was a parallel set of meetings for NGOs from both developed and developing countries, the conference being seen as a key focus for pressure groups active in campaigns connected to development and environmental issues. The conference was held five years after the publication of the Brundtland Report (Chapter 2) to both report and build upon progress. Despite the complexity of the issues under discussion and the divergent interests of the multifarious government representatives and NGO lobbyists UNCED ended with agreed declarations on four major initiatives:

- Agenda 21;
- the Framework Convention on Climate Change (FCCC);
- the Biodiversity Convention;
- Statement of Principles on the Management and Conservation of the World's Forests.

Agenda 21

Agenda 21 is an agenda of action to move towards more sustainable patterns of development in the twenty-first century. With forty chapters and 600 pages the document is very wide ranging and moves beyond issues of principle and policy to deal with proposals for implementation in some detail (UNCED 1993). In short, Agenda 21 aims to be a handbook for sustainable developers, with chapters focusing on specific groups and agencies, be they women, young people, indigenous peoples, trade unionists, scientists and technologists, businesses or local authorities – this list is far from comprehensive but gives an idea of the scope of the document. The interpretation of sustainable development used is broad. Thus the document proposes measures for the conservation of environmental capital, the relief of poverty through economic growth, and the empowerment of disenfranchised peoples.

Since 1992 progress towards implementing Agenda 21 has been slow and limited. Many countries have developed strategies for sustainable development as a response to Agenda 21 and some interesting and worthwhile projects have been initiated as a result. Local authorities, companies and voluntary organisations have followed suit, in both the developed and the developing world. However, at the global level the linked problems of environment and development have got worse, rather than better. To counteract every environmental success story, usually in the developed world (e.g. Box 7.1), there is ample evidence of new or worsening environmental problems. The figures given earlier in this chapter on poverty speak for themselves.

The reasons for this failure are complex and include the difficulties of trade and debt discussed earlier. But the prime reason why Agenda 21 itself has made such small inroads into global poverty and environmental degradation is the lack of funding which was available to implement it. The need for funding was of course foreseen and responsibility for this allocated to the Global Environment Facility (GEF). This was an existing fund, administered by the World Bank, UNDP and UNEP. It was established to assist developing countries and those with post-communist economies to meet the additional costs of responding to global environmental initiatives on climate change, biodiversity, pollution of international waters and ozone depletion. However, the sums available through GEF have always been small when compared with debt repayments made in the opposite direction, from South to North.

The Framework Convention on Climate Change

The Framework Convention on Climate Change (FCCC) bound signatory countries to enter into negotiations aimed at reducing emissions of greenhouse gases. The protracted negotiations to operationalise the FCCC through the Kyoto Protocol are described in Box 2.2. Although the United States and Australia have withdrawn from the FCCC the protocol is expected to come into force during 2003 once it has been ratified by the Russian Federation.

The Biodiversity Convention

This agreement has two distinct themes. First is an agreement between signatory nations to conserve biodiversity at both the species and the ecosystem level and to promote the sustainable use of biological resources. This emerged from long-standing concerns in the industrialised world about the loss of habitats and species, particularly tropical rain forests. The second theme reflects more the agenda of the developing world and seeks to ensure the fair distribution of economic benefits arising the commercial use of from genetic resources.

By the end of 2001 more than 175 countries had ratified the Biodiversity Convention. As with the FCCC, regular Conferences of the Parties (COPs) have been held. Five work programmes have been established for the sectors of marine and coastal biodiversity, agricultural biodiversity, forest biodiversity, the biodiversity of inland waters, and dry and sub-humid lands. Each of these has a vision of, and basic principles to guide, future work; sets out key issues for consideration; identifies potential outputs; and suggests a timetable and means for achieving these outputs. In addition to these programmes 'cross-cutting' initiatives have also been established. As envisaged in the original convention, some issues are relevant to each of the sectors. These have been identified as:

- bio-safety (for example, of genetically manipulated organisms);
- equitable access to genetic resources, traditional knowledge, innovations and practices (particularly where these originate in the developing world but have the potential to be exploited by MNCs);
- intellectual property rights;
- indicators and incentives (see Chapter 8);
- taxonomy;
- public education and awareness;
- problems caused by alien species in particular environments.

Statement of Principles on the Management and Conservation of the World's Forests

As the title makes clear, this agreement was less binding than the hypothetical Forest Convention which had been the original intended outcome of UNCED. Forests were on the agenda of the Earth Summit for the same reasons as biodiversity – concern in the developed world about the erosion of forest cover worldwide but particularly in the tropics. Agreement foundered on the issue of double standards. Developing countries strongly objected to proposed restraints on their use of natural resources within their own borders, pointing out that large parts of Europe had been stripped bare of trees as an essential part of the industrialisation of this region and that unsustainable forestry practices still continued in both North America and Europe. As the United States, Canada and some Scandinavian countries were unwilling to make the financial sacrifices necessary to ensure the sustainable management of temperate forests it was hardly reasonable to expect the much poorer tropical nations to take the necessary action.

Although the Statement of Principles is sensible and far-reaching, referring to social and economic aspects of forests as well as their environmental value, and recognising the need for the involvement of local people in management decisions, no implementation plan was ever adopted. This initiative is now subsumed within the forest biodiversity work programme under the Biodiversity Convention.

The Johannesburg Summit

In 1997 the Earth Summit was reconvened in New York as Earth Summit II but little was achieved beyond the opportunity to note the lack of progress and sense of disappointment from developing nations and environmental groups. A further meeting, the World Summit on Sustainable Development, took place in 2002 in Johannesburg. The starting point at Johannesburg was that the necessary overarching agreements had been put into place ten years previously but both inequality and environmental degradation had worsened since Rio. What was needed was a focus on implementation of strategies to secure sustainable development.

To this end the Summit set new targets, for example:

- to halve the proportion of people without access to basic sanitation by 2015;
- to use and produce chemicals by 2020 in ways that do not lead to significant adverse effects on human health and the environment;
- on an urgent basis to maintain or restore depleted fish stocks to levels that can produce the maximum sustainable yield, where possible by 2015;
- to achieve by 2010 a significant reduction in the rate of loss of biological diversity (WSSD 2002).

These targets were adopted not only by the participating governments but also by NGOs and businesses at the Summit. Over 300 voluntary partnerships were launched to develop projects and disseminate successful local practice in the hope that the direct involvement of developed and developing world organisations in the Summit processes will remove some of the barriers between policy and implementation. However, the Summit failed to secure action on the transfer of renewable energy technologies to developing countries through additional funding for GEF.

Prospects for the future

This chapter started with an environmental success story as a case study, albeit one that related to the developed world (Box 7.1). If progress towards sustainable development is going to prove possible, despite the poor record since Rio, this type of approach must be transferred to the global level, with rich and poor nations co-operating and sharing the costs of action in a more or less equitable way. It must also be extended beyond the environmental sphere to embrace issues of social and economic development, including world trading rules.

One lesson of this chapter has been that economics and the workings of the capitalist system are often prominent when barriers to sustainable development are identified. Chapter 8 therefore examines whether economics must impede progress towards sustainability, or whether it is, in fact, fundamental to achieving sustainable development.

Further reading

Adams (2001) offers a comprehensive overview of the dilemmas that confront attempts to implement theories of sustainable development in developing nations. Desai and Potter (2002) is a substantial collection of essays on the same theme. Vogler (2000) and Francioni (2001) examine the emerging institutions of international governance described in this chapter, whilst in Schuurman (2001) the collected essays review the wide ranging impacts of globalisation on developing nations. Singer (2002) considers some of the ethical issues arising from globalisation. For a stinging critique of the SAP policies of the World Bank and IMF see Stiglitz (2002). This is an accessible and jargon-free text, despite its author being a Nobel Prize-winning former chief economist to the World Bank. A good history of the origins of the debt crisis can be found in Corbridge (1993).

8 Environmental economics

- Supply, demand, price, discounting and externalities
- The economic instruments available to policy makers
- Methods of environmental valuation, including indicators
- Radical approaches to economic theory

Why economics matters

The centrality of economic theory to environmental policy making and sustainable development should be clear from previous chapters. Chapter 1 introduced environmental capital and the economic basis of the definition of resources and of wastes; in Chapter 2 the economic model which is the 'tragedy-of-the-commons' dilemma and the difficulties of valuing the environment and valuing the future were identified; Chapter 3 developed the central role of the economy in providing for human needs and therefore in achieving sustainable development. Economics was a recurring theme in Chapters 5–7 as the policy making cycle at corporate, national and international levels was introduced, ending with consideration of barriers posed by globalisation and international trade to the achievement of global equity and a sustainable future.

Clearly, policy makers need to be conversant with economic concepts, not only understand the basis of environmental problems but also to be able to formulate creative solutions to the recurring problem of meeting human needs in the here-and-now without depleting environmental capital.

Micro-economics and the environment

Micro-economic theory explains how choices by consumer and producers determine the quantity of goods that are produced and the prices at which these are traded. At the heart of the theory lie the laws of supply and demand, which describe the operation of free markets. Free markets occur where participants trade voluntarily and prices are not manipulated by outside parties, for example governments through price controls. Many and competing producers and consumers are another essential feature of free markets. Markets where these conditions apply are said to be examples of perfect competition.

Prices are an essential component of the laws of supply and demand because they convey crucial information to producers, who wish to sell their goods provided they can get an adequate price, and consumers, who may wish to buy if the price is right. This information is shown in graphical form by Figure 8.1. For any given market and for any specific product (good) let us assume that two conditions apply:

● The higher the price, the greater the quantity that producers will be willing to bring to market and sell (this is shown by the supply curve S).
● The lower the price, the greater the quantity that consumers will be prepared to buy (this is shown by the demand curve D).

The point at which the two curves cross is called the equilibrium: at this price (P_e) and quantity (Q_e) there is maximum economic efficiency in the

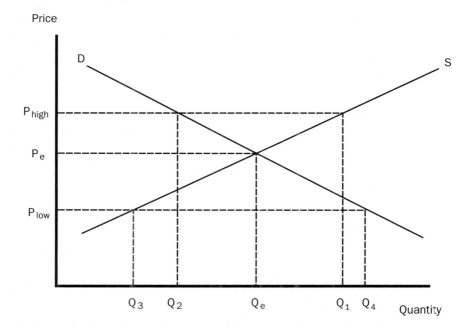

Figure 8.1 *Supply and demand curves*

twin processes of production and consumption. At different prices inefficiencies arise. Either the price is too high (P_{high}) and goods remain unsold ($Q_1 - Q_2$), or the price is too low (P_{low}), all production (Q_3) is consumed, and some consumers are left unsatisfied owing to the higher quantity demanded (Q_4).

Buying and selling under conditions of perfect competition allows producers to adjust prices in response to consumer demand until the equilibrium price is reached. Note that this will only happen if there are many producers in competition with each other. If only a few firms are producing this particular good they might collude to keep the price higher than the equilibrium price, selling fewer units but each at a higher price, ensuring greater profits overall. Competition provides the incentive to lower prices to match those of other producers because over-priced goods may not sell at all.

Of course, changes in market circumstances can change the characteristics and shapes of supply and demand curves. Supply curves depend on the willingness of producers to sell a given quantity at a given price. Producers will become more (or less) willing to sell, and therefore lower (or raise) their unit prices for any given quantity brought to the market, if:

- the cost of inputs (labour, raw materials etc.) changes;
- new technology is developed which lowers the cost of production;
- government regulation, for example a requirement to limit pollution emissions, is made more stringent (and therefore more costly), or relaxed;
- a pollution charge or tax is applied for the first time, increased or reduced.

Figure 8.2 shows that changes which raise the cost of production will move the supply curve upwards (S_1), resulting in a higher price (P_1) and reduced quantity (Q_1) at the equilibrium. Lower production costs will lower the curve (S_2), resulting in a lower price (P_2) and increased quantity (Q_2) at equilibrium.

In Figure 8.3 the results of changes in demand are illustrated. These may be due to:

- changes in the incomes of consumers, leading to increased or decreased willingness to pay;
- changes in taste or fashion which make the goods seem either more or less desirable;
- changes in the price of goods which are substitutes for the goods being marketed. Rice and pasta are examples of goods which are substitutes

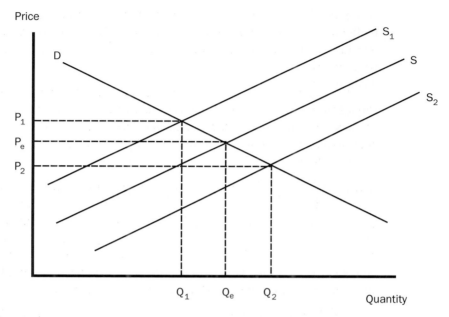

Figure 8.2 The effect of changes in supply characteristics on equilibrium prices and quantities

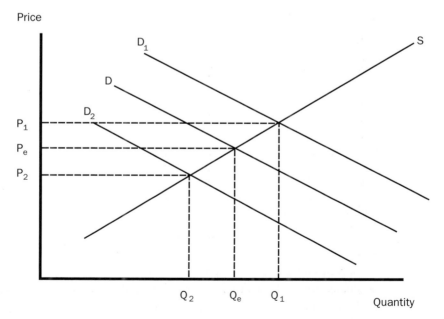

Figure 8.3 The effect of changes in demand characteristics on equilibrium prices and quantities

for each other – if the price of rice increases, some consumers will stop buying it and buy more pasta instead, therefore raising the demand for pasta. This will have the affect of shifting the demand curve for pasta upwards.

Figure 8.3 shows that changes which raise the desirability of goods will move the demand curve upwards (D_1), resulting in a higher price (P_1) and increased quantity (Q_1) at the equilibrium. Reduced desirability will lower the curve (D_2), resulting in a lower price (P_2) and decreased quantity (Q_2) at equilibrium.

The slope of demand curves will vary according to consumer behaviour and attitudes. There are some goods for which quite large increases in price will have little effect on the quantity demanded (see Figure 8.4). When this occurs the demand curve is said to be price-inelastic. By contrast, price-elastic demand curves show large changes in quantity demanded at different prices (as in Figures 8.1–3). A good example of a product for which demand is relatively inelastic is petrol.

How do the operation of free markets and the laws of supply and demand relate to environmental policy and sustainable development? Certainly, market forces lead to resource efficiency by matching supply to demand and forcing efficiency upon competing producers. However, markets on

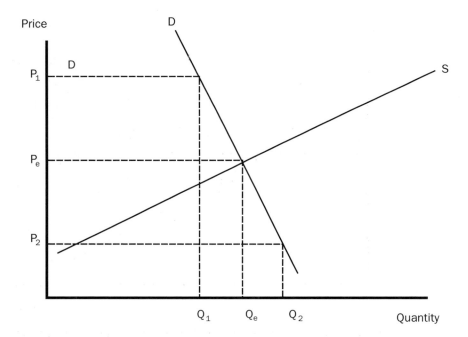

Figure 8.4 Price inelasticity of demand

their own will not achieve outcomes that are compatible with sustainable development. The reasons for this become clear when considering the key criteria for sustainability: equity, futurity and valuing the environment.

In the narrow economic terms of models based on perfect competition, efficiency has nothing to do with equity. Social justice is not even incidental to the outcome. Demand curves only register the willingness to pay of those with sufficient money to join in the trading and so the needs of the poor count for little or nothing. Those able to supply to the market skills that are in high demand will command high wages and high spending power: those with less developed skills, or who are unable to work due to disability, will remain poor. Over time, this wealth discrepancy is exacerbated as those with higher incomes save, turning income into capital, which can then be invested to produce even more income.

Similarly, markets are neglectful of the future. Imagine a world with a zero rate of inflation and suppose you were to be offered £100 sterling, with the choice of accepting it immediately or in one year's time. Even though you knew the value of the money would not be eroded by inflation over the next twelve months you would almost certainly ask to be given the money at once. This would be for two reasons:

- *Time preference*: that is, the desire to have things sooner rather than later. During phases of economic growth, this impatience is compounded by the belief that, because the future will be more affluent, the £100 will be of less significance then than it is now.
- *Productivity of capital*: having the money now allows it to be invested over the next twelve months to produce more wealth.

Money in the future, therefore, has a lower value than money in the here-and-now. How much lower depends on the discount rate applied. Discounting is the technique used by economists to allow for the lower future value of money when assessing the viability of investment

Table 8.1 *The future value of £100 at different discount rates (£)*

	Year				
%	Now (year 0)	1	2	10	50
1	100	101	102	110	164
2	100	102	104	122	269
5	100	105	110	163	1,147
10	100	110	121	259	11,739

decisions which will result in costs and benefits over a period of years. Discount rates work as reverse interest rates and are compounded year on year (Table 8.1). The higher the discount rate and the longer the time period, the larger must be the future expected returns to tempt investors.

Another way of expressing this (and the way most commonly used by economists) is to say that the higher the discount rate used, the lower the present value of both costs and benefits in the future. Present value is simply the value here-and-now of a sum of money in the future once the compound discount rate has been applied. The formula for present value is

$$PV = V_t / (1 + r)^t$$

where PV is present value in year zero, V_t is value in year t, and r = discount rate.

Table 8.2 shows the present value of £100 in future years for different discount rates. Net present value (NPV) is the sum of all discounted costs and benefits during a given time period. A proposed nuclear power station (Box 8.1) would typically be appraised over forty years, but shorter time periods are more usual. Investments are deemed worthwhile in economic terms if the NPV is positive at the required discount rate. Where future returns are uncertain higher discount rates tend to be used, to reward risk taking by investors.

Discounting has unfortunate consequences when used to appraise projects with large costs or benefits in the far future such as the Severn barrage (Box 4.3) or nuclear power stations (Box 8.1). The general case is that long-term costs (such as expensive decommissioning of plant and equipment) and long-term benefits (for example the mature forest that will result in fifty years time from a decision to plant trees now) are both diminished in value when discount rates are used by decision makers.

Table 8.2 The present value of £100 in future years for a range of discount rates (£)

	Value in year 0 of £100 in year			
Discount rate (%)	1	2	10	50
1	99	98	91	61
2	98	96	82	37
5	95	91	61	9
10	91	83	39	1

Box 8.1

Balancing inter-generational costs: US nuclear power

Nuclear power stations have high capital costs because they are expensive to build. Significant costs also arise during the decommissioning phase, which can extend to up to 200 years after the station was constructed. In addition, some nuclear wastes have the potential to cause environmental damage thousands of years into the future.

Policy making for radioactive waste management in the United States has attempted to balance inter-generational costs and benefits. Particular problems have emerged with the disposal of high-level waste (HLW), which is mostly spent fuel rods removed from nuclear power stations. This waste needs to be stored under special conditions and can be disposed of only into a specially engineered repository, deep underground.

The final disposal of HLW is a federal government responsibility and in 1982 the Nuclear Waste Policy Act was passed. This gives responsibility for the research and development of HLW disposal facilities to the US Department of Energy. Such development was funded by a levy on electricity generated by nuclear power, thus ensuring electricity consumers were making a contribution to the project.

The Act specifies that for the first 10,000 years (i.e. until AD 12000) the risk of premature cancer deaths from the repository must be less than 1000 lives. The decay of radioactivity within the waste can be precisely calculated, provided the initial level of radioactivity and the radiochemical composition of the material are known. Less predictable are the geological and hydrological characteristics of the repository site. Characterisation work at the Yucca Mountain site over a decade or so has been undertaken so that computer models can be developed to predict whether the repository will meet the tough standards demanded by the 1982 Act. In July 2002 the project gained presidential approval, on the basis that the Yucca Mountain site is scientifically sound and suitable for development as the nation's long-term geological repository for nuclear waste.

The behaviour of people in the future with respect to the site is completely unpredictable. The Act states that the repository must be designed so that the waste is protected from human intrusion by multiple barriers, and by markers which signify danger in a non-linguistic way, yet waste must also be retrievable, so that if unforeseen problems emerge, any necessary action can be taken. Therefore there is a possibility that in the future individuals, by accident or design, may harm themselves or others by deliberately bringing waste out of the repository and into contact with human beings. It is certain that the cost of guarding the waste will be considerable.

References: Blowers *et al.* (1991), especially chapter 5; Office of Civilian Radioactive Waste Management (2003).

Box 8.1 continued

Discussion points

1 If one objective of US policy has been a fairer distribution of costs and benefits between present-day and future stakeholders, how successfully has this been achieved?

2 When appraising a proposed nuclear power station, what would be the effect of raising or lowering discount rates on the present value of (a) *electricity revenue* (anticipated in, say, years 5 to 45), (b) *decommissioning costs* (incurred in, say, years 46 to 200, with a peak in year 200 when the final demolition of the station occurs), (c) *waste management* costs in AD 12000?

This diminution increases as larger discount rates are used and as the time period increases. The effect is built-in discrimination in favour of the present at the expense of the future.

Some philosophers (Goodin 1982; Parfit 1983) have argued that the only discount rate which is fair to future generations is 0 per cent or even negative – giving equal or higher weighting to future costs than to present ones. They reinforce this argument by pointing out that the assumption that the future will be more affluent is dubious, given population growth and the rate at which the ability of the environment to sustain the economy is being undermined. The problem with this argument is that low discount rates tend to encourage investment in capital infrastructure such as factories, power stations and roads, and therefore the increased depletion of resources and the production of wastes. Table 8.1 shows why an investment of £100 would have to yield only £110 over ten years to be deemed justifiable at 1 per cent discount rate, but £259 yield would be required to justify it at 10 per cent. Higher discount rates mean only the most profitable investments will proceed.

So, market forces alone cannot protect the future. What about the third sustainability criterion, valuing the environment? The problem which immediately arises is that the environment consists of a set of interacting systems, most of which cannot be bought and sold. Parts of the environment can be traded – land, for example, or rights to fish in a stretch of river. But most aspects of the environment are open-access resources – a concept first introduced in Chapter 2 when the tragedy of the commons was discussed. Because the use of the atmosphere as a sink, for example, is not traded, the laws of supply and demand cannot regulate its use via the price mechanism. It is free for all to use and consequently will tend to be over-used, just as the common land in the tragedy example becomes over-grazed. As well as being environmentally damaging this is also economically inefficient.

Unpriced costs and benefits are called externalities (or external costs and external benefits). Beekeeping is an example of an individual activity which gives rise to both types of externality. The cost of maintaining the hives and the profit from the honey all remain with the beekeeper as internal costs. But neighbouring gardeners will benefit from the pollination of fruit and vegetables by the visiting bees – but may also have to put up with the occasional sting. Social costs are the sum of internal and external costs, that is the total costs across all affected parties, of a particular set of activities.

The 'polluter pays' principle in practice

Economic instruments

Economic instruments were introduced in Chapter 6 as a means of using the price mechanism to implement environmental policies. Economic instruments have the effect of internalising external costs – making the polluter bear the cost of the environmental damage. This is the practical application of the 'polluter pays' principle. Bringing externalities into the cost of product by imposing a tax (t) will shift the supply curve (Figure 8.5), resulting in lower overall quantities once a new equilibrium (P_1, Q_1) is established on the basis of social, rather than just internal, costs. The principle is well established internationally, having been adopted as a basis for international trade by Organisation of Economic Co-operation and Development (OECD) in 1975.

Common objections to the polluter pays principle are that it is distasteful to trade in environmental degradation and also to allow those who can afford it to pollute the environment of others (Pepper 1996: 77–80). Both of these are relevant to the discussion above of the incompatibility of free markets and progress towards sustainable development. Valuing the environment is inevitably contentious: beyond the technical issues outlined later in this chapter there are the ethical arguments about intrinsic and extrinsic values which were introduced in Chapter 2. Two arguments can be brought against the first objection. Forcing polluters to pay for the environmental costs they impose on others is far preferable to them treating the environment as though it was worthless. Licensing pollution (which is the regulatory alternative to economic instruments)

Price

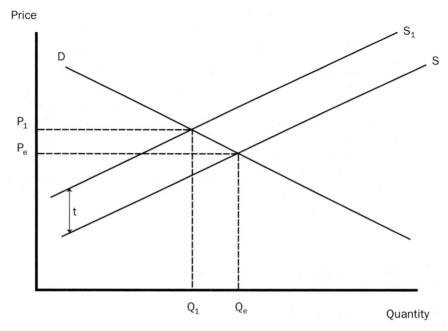

Figure 8.5 *The effect of a pollution tax, t, on price and quantity*

should logically be as distasteful as trading it. The second objection has greater force. Economic instruments operating in competitive markets may produce outcomes that are 'efficient' in narrow economic terms; however, these outcomes may not necessarily be fair.

Pollution charges and tradable permits

When applying the 'polluter pays' principle there are several types of economic instruments to choose from and each is suitable in different situations. All usually require legislation to enforce them if free-rider problems are to be avoided.

Pollution charging, first introduced in Chapter 6, is a tax on gaseous, liquid or solid wastes. This instrument is suitable for large, fixed sources of emissions such as factories and power stations (Figure 6.4). Pollution is measured and charges are paid per unit produced. The regulatory agency has a key role in negotiating the level of the charge and providing independent monitoring of emissions levels to ensure compliance. The UK Landfill Tax introduced in Box 8.2 is an example of a pollution tax.

Tradable pollution permits are a more sophisticated form of pollution charging. Under this system, which was piloted in the United States as a

Box 8.2

The UK limestone and landfill taxes

In the United Kingdom limestone from quarries, some of which are in environmentally sensitive areas, is routinely used for low-grade purposes such as road-building aggregate. At the same time, demolition waste, some of which could have replaced the virgin limestone, is instead being sent in large quantities to landfill sites, themselves a rapidly diminishing resource in some parts of the country. Recently introduced landfill and aggregate taxes have the potential to decrease this throughput by raising the costs of extraction and disposal, thereby making recycling and reuse more commercially attractive.

A landfill tax was introduced in 1996 with the aims of:

- internalising the external costs of landfill;
- increasing recycling and valorisation of waste;
- reducing the volumes sent to landfill (ECOTEC 1998).

The tax was set at a dual rate. Inert wastes, including demolition wastes as well as rocks, soil, concrete etc., were taxed at £2 per tonne. Active wastes (defined as those which would decompose in landfill) were initially taxed at £7 per tonne. Revenues are partly hypothecated to locally based environmental trusts for environmental improvement schemes in the vicinity of landfill sites.

The initial rates were based on a valuation study (DoE 1993) of the external costs of landfill. Subsequent increases in the rate for active wastes (due to reach £15 per tonne in April 2004) put the tax well above the original valuations. This can be justified in economic terms because the original study explicitly stated that it had been unable to value some significant externalities. The external benefits of waste minimisation and recovery were also not allowed for in the study. But it has been argued that, given the limitations inherent in environmental valuation, it would be legitimate to set a tax rate with reference only to desired behaviour changes (the second and third aims) and without the use of environmental valuation (ECOTEC 1998).

Between 1996 and 2002 the volume of inactive waste sent to landfill decreased by 60 per cent, but the volume of active waste has not changed significantly (Strategy Unit 2002). This difference is in part due to the higher than average percentage of waste expenditure to turnover in the construction industry. Inactive waste diverted from landfill is now being recycled into construction and landscaping projects, although a small amount may be being disposed of illegally (ECOTEC 1998).

The Aggregates Levy, introduced in 2002, will complement the effects of the landfill tax by raising the cost of virgin aggregate by £1.60 per tonne. The aims of the levy are:

- to internalise the externalities incurred by aggregate production;
- to encourage substitution and recycling;
- to promote the more efficient use of primary aggregate

(EDS 2001)

Box 8.2 continued

Although the slate and china clay industries foresee new markets for their waste as substitutes for primary materials, the possible extent of further substitution is uncertain. Potential barriers include the inconsistent quality of reclaimed materials and possible contamination of some sources, for example furnace ash from incinerators. The development of quality specifications could help to overcome these difficulties.

The effect of the two policy measures taken together will decrease throughput by deterring the extraction of raw material and final disposal of wastes. However, there are externalities associated with reclamation activities, such as noise, dust and transport nuisance, which remain unaccounted for and may distort the overall environmental benefits of these measures.

References: ECOTEC Research and Consulting (1998); EDS (2001); Strategy Unit (2002).

Discussion point

1   Waste minimisation is the highest level of the waste management hierarchy (Figure 4.3). To what extent will the landfill tax and Aggregates Levy encourage minimisation, for example by making developers choose to restore existing buildings rather than demolish them? Consider the taxes at their existing levels and the effect that a substantial increase might have.

means of controlling sulphur emissions from large plant, factories can trade emissions permits between themselves. Those who find it cost-effective to reduce emissions will have surplus permits, which can be sold to competitors who are less able to cut their pollution. For example, coastal power stations might import low-sulphur fuels and sell their excess permits to stations far inland, which have less choice of fuel. The regulator can issue permits in one of two ways. *Grandfathering* recognises the rights of established industries, which are granted permits based on their emission levels prior to the introduction of the trading system. The alternative is to auction permits each year to the highest bidders.

The advantage of the tradable permit system over simple pollution charging is that it allows the regulator precisely to ratchet down, year by year, the total amount of pollution produced from large plant by steadily decreasing the number of permits in the system. However, if the pollutant has a high local environmental impact tradable permits may not be appropriate. The operation of the permit market might lead to emissions being concentrated in relatively few geographical areas. Alternatively, plants whose locations mitigate the effects of local pollution (for example power stations to the windward side of acid-tolerant limestone landscapes) may reduce their emissions and sell their spare permits to plants downwind of more vulnerable soils, resulting in small or even negative net environmental benefit.

Product charges

Charging polluters directly, whether by pollution taxes or tradable permits, will raise the price of production and some of the increase will be passed on to consumers. This does not contradict the 'polluter pays' principle – it is right that the consumers of products should pay the full social cost of manufacture. Sometimes the bulk of pollution during a product's life cycle occurs not during the manufacturing stage but during or after its use by consumers. A good example here is transport fuels but many other products fall into this category, especially if they pose particular disposal difficulties at the end of their life such as vehicle tyres and batteries containing heavy metals.

Emission charging is obviously impracticable and unenforceable in these cases. The task of monitoring emissions (and collecting the due taxes in arrears) from drivers of the several million vehicle exhaust pipes that pump out pollutants each day in every developed and developing country would defeat the most assiduous regulatory body. A much simpler option is chosen: petrol and diesel for transport use are taxed at the pump. Product charges make the polluter pay up-front.

Product charges may take the form of returnable deposits for wastes that are potentially recyclable or reusable. Few people born after 1965 in the United Kingdom will remember how ubiquitous the deposit-refund scheme for glass drink containers used to be. Increasing affluence, competition from other packaging materials such as aluminium and plastic, and the strenuous resistance of the emerging supermarket chains to handling returned bottles saw the practice become nearly obsolete in some countries, although where enforced by legislation (as for example in some US states and Finland, Jacobs 1991: 142) it remains an effective policy instrument.

Resource taxes

Resource charges move the point of taxation even further back in the resource cycle (Figures 1.1–1.2) – not to the point of emission, not to the point of sale to the final consumer, but to the beginning of the resource cycle. On the face of it this is a fundamental approach to limiting environmental damage, building in automatic producer responsibility from the outset. Resource taxes are appropriate particularly in cases where recycled alternatives to virgin resources offer environmental benefits but little or no economic advantage. A good example is the UK tax on limestone extraction introduced in Box 8.2.

But what is the polluter paying for?

The analysis thus far of economic instruments has focused on the narrow effects of their use at the point in the resource cycle at which they are applied. However, what are the uses to which money raised by pollution charging can be put?

In all cases some of the money raised will pass to the regulatory body simply to cover its costs in enforcing the pollution control regime. Where there is a surplus, two broad options exist. Governments (at national or local level) can add the revenue to their general taxation revenues, either to fund additional non-related expenditure or, more usually, to reduce the level of tax on other activities such as earnings or employment. Alternatively, the revenue can be applied to a specific purpose in order to mitigate the environmental problem caused by the polluting activity. The earmarking of tax revenue for any purpose is called *hypothecation*. One straightforward example is the use of UK Landfill Tax revenue for environmental enhancement projects in the vicinity of landfill sites (Box 8.2).

Hypothecated revenue can also be used to increase the effectiveness of pollution taxes by subsidising alternatives to the product or activity causing the pollution. Any local recycling projects funded by Landfill Tax revenue would be examples here; so is the use of a levy on fossil-generated electricity to support renewable energy projects (Box 6.2). Hypothecation is particularly appropriate where the demand for the polluting activity or product is inelastic (such as transport fuels) and makes a significant contribution to the quality of life of the poor. Using petrol taxation to price the middle classes out of their cars and on to buses may be fair: but the same tax might affect severely those on low incomes (especially in rural areas) who are dependent on their vehicles to get to work and to access other essential services. If the buses do not exist or are expensive to use one obvious use for the revenues is to subsidise public transport. Providing a substitute good (buses) for an essential service (mobility) has two effects on the demand curve for petrol (Figure 8.6). The curve shifts downwards (from D to D_1), reflecting the reduced

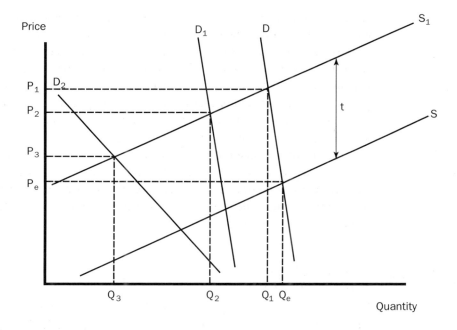

Figure 8.6 *The effect of a hypothecated pollution tax on price and quantity*

desirability of petrol once a substitute exists. Over time it also reduces in gradient (from D_1 to D_2), that is becomes more elastic, as private motoring becomes less of an essential and more of a luxury good. These shifts increase the effectiveness of the tax (t) in reducing demand. The initial effect (Q_e–Q_1) is small but the medium-term (Q_e–Q_2) and longer-term (Q_e–Q_3) reductions are much larger.

Subsidies and government expenditure

Subsidies to reward changes in behaviour are of course yet another type of economic instrument. Although such expenditure may be hypothecated it has been more common for governments to subsidise certain activities out of the general pool of taxation revenue. Grants to farmers to manage land for biodiversity as well as agricultural production, grants to householders for part of the cost of thermal insulation material and grants to motorists who convert their vehicles from petrol to gas-based fuels are all examples of ways in which government expenditure has been used to change behaviour and reduce environmental damage.

However, subsidies are rarely the first choice of economic instrument. Unless funded by hypothecated revenue they breach the 'polluter pays' principle because the community pays the polluter, rather than the other way around. Another fundamental problem is the inherent inefficiency of subsidies. In the case of domestic insulation, for example, a certain proportion of householders would choose to insulate their homes without any subsidy at all, but will claim the subsidy if it is available. A further proportion would make the change for significantly less money than is in fact on offer. Subsidy payments in these cases are wholly or partially wasted. Subsidies to companies in return for pollution cuts suffer from the same drawback. In addition, in the long run they may result in more pollution overall. This is because, although subsidies encourage production and pollution cuts by individual firms, they lower the average cost of production, thus encouraging more competitors to enter the market.

Valuing the environment

The concept that the polluter pays is both conceptually attractive and, through the use of economic instruments, practically useful for those seeking to implement environmental policy. However, difficulties in valuing external costs and benefits can limit the scope of economic instruments. It has already been noted that externalities are goods (and bads) for which there are no markets. Such goods are called public goods: air quality, for example, whether good or bad, is enjoyed or endured by all who live in a particular location. However, without markets there will be no prices to indicate how much good air quality is worth and therefore how much the polluter should pay.

Valuation methods have been developed so that unpriced goods can be valued in money terms. This is potentially of very great importance to policy making for sustainability because valuing the environment is an essential component of decision making for sustainable development. Moreover, in Chapter 3 it was noted that the environmental capital approach, combined with environmental valuation techniques, offers a way of measuring progress towards sustainability. However, before describing these techniques it is important to recognise some of their inherent limitations:

● Environmental valuation is contested by eco-centrist thinking which holds that it is not morally acceptable to reduce unpriced environmental capital and services to numbers on a balance sheet. Money-based values can only ever represent extrinsic value (Chapter 2). This means that decisions made on the basis of these techniques may be contested on ethical grounds.

- There are significant methodological issues about the meaningfulness of some of the environmental valuation techniques.
- Policy makers and those seeking to influence them will tend to seize on the apparently solid information which monetary valuation provides. But the techniques can only ever provide an approximate value and very large ranges of uncertainty are often given for particular valuations.
- All the techniques are seeking, through research, to measure human preferences and translate these into monetary values. The preferences of future generations, however, cannot be measured and so have to be assumed.
- Environmental valuation techniques have the advantage of discovering the strength of individual preferences – but they also tend to undervalue the preferences of lower income groups.
- Many aspects of environmental quality cannot be valued in money terms to any meaningful extent. This can be for many reasons but is usually due to the complexity of environmental systems and lack of knowledge about the effects of environmental changes, especially in the long term.

Revealed preference methods

When consumers spend money on traded goods and services they are expressing their preferences. The aggregated demands of all consumers in any particular market constitute the demand curve (see Figures 8.1–6). For some non-traded goods research data on consumer behaviour in other markets can be used to derive an estimated demand curve by revealing consumer preferences.

The travel cost method can be used to estimate the recreational value of sites. Visitors are asked how often they visit and how much they have paid to make the journey and spent at the site. The resulting data can be used to derive demand for the site. Hedonic pricing uses property prices to infer estimates of the value of aspects of environmental quality. A house overlooking a busy motorway will sell for less than an identical dwelling built away from the visual and noise impacts of the road. Comparing the prices at which houses are bought and sold will reveal the price of these impacts. Of course, such surveys are complicated by the need to allow for all the other factors which affect house prices, including the design of the dwelling and its garden, its location relative to local services and other aspects of environmental quality. For example, if hedonic pricing was being used to determine the cost of motorway nuisance, the proximity of industry, which is another factor likely to lower house prices, would have to be allowed for in the calculations. Such technical difficulties mean that hedonic pricing surveys have to be designed very carefully and that they will be suitable only in situations where available data are sufficient in quantity and quality.

Stated preference methods

Revealed preference modes rely on the availability of data about consumer demand for activities related to the cost or benefit being valued. Some environmental assets, however, are widely valued even though they are never visited or directly experienced by the majority of people. In these cases travel cost or hedonic pricing data will not exist. However, it may be possible to undertake a survey of stated preferences by contingent valuation. This is a widely used method of environmental valuation with an extensive academic literature.

Contingent valuation studies survey a representative sample of a given population about hypothetical markets – that is, how much they theoretically would be willing to pay (WTP) in order to preserve a particular environmental asset, or willing to accept (WTA) to allow its deterioration. The demand for the asset can then be estimated in terms of the whole population. When designing contingent valuation studies decisions have to be made about the following issues, all of which can potentially affect the resulting valuation:

- the amount, style and content of information provided to the interviewee prior to interview about the asset being valued;
- the hypothetical means of payment – charitable donation, extra taxation, or fee to visit the asset;
- will the WTP questions be open (how much would you pay?) or closed (would you pay £x?);
- if open questions are asked, how to allow for the 'free rider' effect by which those who feel strongly about the preservation of the asset offer very large sums knowing they will never be called upon to pay.

Cost–benefit analysis

Deriving valuations of environmental attributes is only useful to the extent that these can be fed into the decision making process. For project proposals the most usual way to do this is through the process of cost–benefit analysis (CBA). This technique involves calculating the economic value of all the costs and benefits (both internal and external) which are anticipated to arise from a particular proposal. An appropriate discount rate is then chosen and the net present value (NPV) (that is, the sum of all discounted costs and benefits during a given time period) of the project is calculated. If NPV is positive the calculated benefits outweigh the costs – this means the project is worth while. The formula for NPV is

$$NPV = \sum_t \frac{B_t - C_t}{(1 + r)^t}$$

where B_t = total benefits in year t, C_t = total costs in year t and r = discount rate.

CBA is much used by policy makers, especially in the public sector, to bring market and non-market costs and benefits into a common framework of analysis. As well as environmental costs, other externalities such as the value of the travel time of commuters and the costs of road traffic accidents either engendered or avoided by the proposal can be fed into the analysis.

Caution needs to be exercised, however, when using the results of CBA in decision making. For projects with effects predicted to reach into the medium to long-term future, discounting may disadvantage future generations relative to the present, as discussed above. CBA is based on predicted costs and benefits and therefore always subject to a degree of uncertainty. Unanticipated economic circumstances may increase or reduce the costs and benefits of the project; as may changes in public attitudes if they change the demand for the services produced by the project or lead to more stringent regulation of the activity. Where the activity is novel and likely to produce environmental change, the ramifications of this may be poorly understood and therefore costs and benefits impossible to estimate in a meaningful way. Finally, CBA can incorporate only those attributes of the environment for which it is possible to derive values – some costs and benefits will always remain external to the equation. CBA therefore produces significant information to feed into the political decision-making processes described in Chapter 6 – but it is not a substitute for them.

Measuring sustainable development

The previous sections have demonstrated the utility of money as a measure of some aspects of environmental quality but also its limitations. To what extent can money measure quality of life? Traditionally, politicians and economists have made the assumption that living standards in any country can be summarised by statistical indicators based on economic performance. The headline indicators used are Gross Domestic Product (GDP) and Gross National Product (GNP). Both of these are calculated by totalling the economic output of a country in a given time period. GDP includes all economic activity within the nation regardless of ownership whilst GNP excludes production owned by non-residents but includes activity carried out abroad under the ownership of those resident in the country.

Steady growth in GNP is the goal of most governments in the developed world and the developing world. When the economy is growing, more wealth is available to the population and, provided the proportional increase is available to most of the population, people in general will feel wealthier. When economic growth falters, even to a small extent or for a relatively short period, the effects are noticeable in people's daily lives. Unemployment rises so that even those in work may worry about their medium-term prospects. This can lead to a further economic downturn, as consumers become more cautious, saving rather than spending their money. In severe recessions the value of assets such as houses and shares can start to fall. Democratically elected politicians have a vested interest in growth.

For environmental economics growth is problematic, for two quite separate reasons. First, and fundamentally, the sustainability of long-term economic growth is questionable if exponential growth means exponential increases in resource utilisation and waste production. The second issue concerns the extent to which growth in GNP can be equated with increases in quality of life. To the extent that wealth contributes to quality of life the relationship is straightforward. However, if people choose to work less overtime because they would prefer the leisure, this might improve their quality of life. GNP, however, would fall because less work means less production and less consumption if wages are forgone. Moreover, some forms of economic growth correlate with decreases in quality of life. The aftermath of earthquakes, car accidents, plane crashes or any other type of disaster will produce plenty of economic activity as medical, funeral, repair and replacement activities make work for different agencies and money changes hands. GNP will rise but the overall quality of life is diminished overall.

The issue of external costs also arises. GNP measures only what is traded and cannot account for environmental externalities such as pollution, nor social externalities such as fear of crime, although these may have a severe impact on quality of life. Neither are external benefits counted. Improvements in air quality may come at the price of lower output – this will have a negative effect on GNP but the positive effect simply is not counted. Voluntary work in the community and labour within families such as child care make a massive contribution to society which GDP does not recognise at all.

One further issue is the depletion of capital, both human-made and natural. The 'gross' in GNP denotes that no allowance has been made for the depreciation of capital assets. When estimated capital depreciation at a national level is calculated and deducted from GDP the resulting indicator is called the Net National Product (NNP). However, only human-made capital is included in this calculation. This exclusion of natural capital from GNP and NNP diminishes further the usefulness of these indicators as measures of sustainable wealth creation but is of particular significance in developing countries. These are often dependent on exports of unprocessed commodities such as timber and agricultural goods whose production may deplete environmental capital very directly. Loss of biodiversity, forest depletion, soil erosion and lowered fertility due to pollution from agri-chemicals are examples of loss of natural capital, unsustainable in the long term, but invisible if GDP/NNP are used as the measures of wealth. Table 8.3 shows estimates of some types of

Table 8.3 *The cost of environmental degradation in Nigeria (US$ million/year)*

Type of degradation	Annual cost of inaction
Soil degradation	3,000
Water contamination	1,000
Deforestation	750
Coastal erosion	150
Gully erosion	100
Fishery losses	50
Water hyacinth	50
Wildlife losses	10
Total	5,110

Source: World Bank (1990), quoted in UNEP (2000b)

natural capital depletion in Nigeria, totalling more than 15 per cent of GNP (UNEP 2000b).

It is obviously important to be able to measure progress towards sustainable development, but GNP and its related indicators are unable to do so. There are two potential solutions to this problem. One is to make adjustments to GNP so that the measure starts to incorporate external costs and benefits, whether these are transient (such as noise pollution) or longer term enhancements (or depletions) of natural capital. Deductions from GNP can be made to allow for expenditure that is remedial in nature, such as dealing with the aftermath of accidents or cleaning up environmental pollution. In this way GNP would become a more reliable indicator of the extent of positive or negative changes in human welfare year on year. Some such measures are in use: Japan pioneered national environmental accounting along similar lines in the 1970s and other countries, such as Sweden, Norway and France, have followed. Offsetting internal and environmental costs in this way, of course, measures weak sustainability (Figure 3.3). Rapid growth in the economy could mask irreversible and detrimental environmental damage, prejudicing the interests of future generations.

The other potential solution is to accept GNP as what it is, a limited measure of money costs and benefits, and develop complementary measures (indicators) of social and environmental externalities. Such indicators will not necessarily be expressed in monetary terms. However, just as conventional accounting differentiates between capital and revenue, so environmental indicators must distinguish the changes in stock and flow resources and waste streams. Sustainability indicators must also include measures of social changes affecting equity. Table 8.4 shows the indicators of the quality and sustainability of urban environments that have been developed by the United Nations Environment Programme (UNEP). Note that 'city product' is the city-scale equivalent of GNP, but that this is but one of the measures contributing to the overall assessment.

Indicators have the disadvantage, when compared with methods that put monetary values on to externalities, that the relative importance of the figures generated is not explicit. Their advantage is that each factor is identified clearly and not rolled into a 'headline' figure, which could mask some important features of the overall position. A further advantage is that, because environmental features are accounted for separately, it is possible to monitor for strong sustainability using indicators by applying the principle of constant environmental capital.

Table 8.4 *Indicators of the quality and sustainability of urban environments*

Access to drinking water	Poor households
Air emissions	Population density
Air quality	Population growth
City product	Presence of LA 21 process
Energy consumption	Price of water
Green areas	Quality of drinking water
Health care	Recycling
Housing price	Rent-to-income ratio
Infant mortality	Safety
Investments in green areas	School attendance
Investments in water supply systems	Transport modes
Organisations using environmental audit systems	Travel times
	Waste production
Participation in decision making	Waste water treatment
Participation in elections	Water consumption

Source: CEROI (2001)

Note: Information on how each indicator is derived can be found on the CEROI Web site (CEROI 2001).

A new economics?

Economic activity is intimately connected with sustainable development. Equity can only be achieved through the creation of wealth and its distribution to those presently in poverty; however, the natural capital entitlement of future generations can only be protected by forms of economic activity which limit their environmental impact to that which is sustainable. Economic science seeks to describe the operations of markets (positive economics) – but it is also often used to prescribe how those markets should be run (normative economics). During the twentieth century the key debates were those between the different prescriptions of Marxism, socialism, Keynsianism and monetarism. This century, the ideas of environmental economics and ecological economics are likely to be as influential.

Whereas conventional economics recognises externalities but focuses its main attention on internal transactions in the economy, environmental economics is the straightforward application of the ideas introduced in this chapter to internalise and account for full social costs and benefits. Following the 1992 Earth Summit, environmental economics has grown in significance across the world as governments at international, national and the local levels attempt to use its insights to develop and implement policies to achieve a more sustainable pattern of growth. Environmental economics is undoubtedly normative because its fundamental premise is that the environment matters. When considering issues of sustainability, environmental economists will have a presumption in favour of equity and will incorporate much longer time periods into their analyses than would be usual for conventional economists.

Ecological economics, however, goes much further in its prescriptions (Faber *et al*. 1996). Intrinsic value is a fundamental tenet of ecological economics. Human preferences are not the sole source of value: environmental attributes have value and this cannot be expressed in terms of money – indeed, to do so is to devalue them as by implication they can then be damaged or destroyed in return for a given sum of money. Ecological economists attempt to combine ecocentric principles with economic theory in order to advocate large-scale changes in the current economic order.

Ecological economics rejects the conventional analysis of globalisation as leading to greater economic prosperity for all. Local and small-scale economic activity is preferred because, it is claimed, this is more likely to meet the needs of local populations. These needs are not just for consumer goods, but also for environmental quality; meaningful work; and a strong community. If communities are producing goods for local consumption transport externalities are avoided. Those who are dependent upon the local environment will manage it responsibly. Work for a small firm producing goods for one's own community is more likely to give satisfaction than is alienating labour in a mass-production factory for a multinational company. Economic interdependence at the local level will foster a healthy community spirit and thus enhance the welfare of all. One application of these ideas is explored in Box 8.3, a case study of Local Exchange Trading Systems (LETS).

One further important feature of ecological economics is its emphasis on the gap between the ideal-typical models used by conventional economics and what happens when real people trade in real markets. Very large assumptions are necessary to frame even simple economic models, such as those illustrated in the figures in this chapter. For example, the laws of

Box 8.3

Local exchange trading systems

Local exchange trading systems (LETS) allow the trading of goods and services within local communities using currencies issued and controlled by the trading group. Although they are often set at a rate which is related to the official currency and therefore potentially convertible, LETS currencies differ from money in several important respects. LETS schemes are co-operative and non-profit-making; all transactions are centrally recorded (there are no notes or tokens to trade with); information about members' balances is available to all; the LETS currency can be used only to buy goods and services, so it cannot be invested, borrowed or earn interest.

By early 2003 there were 40,000 people involved in 450 LETS schemes in the United Kingdom (Letslink 2003), with other schemes set up in Canada (where the idea originated), the United States and Australia. The primary attraction of the scheme is that it allows trading to take place without money. Mrs A. is retired and needs some help with gardening but cannot afford to pay a gardener. She babysits for Ms B. and is paid in LETS units. This allows her to pay Mr C. for gardening work. Mr C. spends his units on training in computing skills, offered by Ms D. Some of this activity might have happened anyway through the exchange of favours between friends. LETS, however, allows such exchange between strangers and stimulates activity by removing the sense of obligation involved in asking for favours.

At any stage, some members will be in credit and some in debit, because the net number of units will be zero. The scheme cannot work unless some members have accepted more goods and services from others than they have yet contributed. To have a debt to the system is called being 'in commitment'. People in commitment could choose to leave the scheme without offering goods or services to clear their debt, or may die. This is less of a problem than it might appear. By running up debts they have credited other scheme members, who will be able to spend their units buying goods and services from other members. However, a member leaving with small debts to a large number of members might result in reduced trading activity, as those with credit become reluctant to offer more work in to the system. Central recording of transactions and the transparency of members' accounts mean that members could refuse to offer further goods and services to someone who is exploiting the system.

LETS systems claim economic, social and environmental benefits. The increased activity improves the welfare of those doing the work and those benefiting from it. Social inclusion is fostered and communities are strengthened through the increased interaction that the scheme fosters. The environment benefits if goods can be repaired rather than replaced; if expensive tools and equipment can be shared rather than bought by every household; and through the reduced transport externalities associated with sourcing goods and services locally.

Difficulties and pitfalls include:

- ambiguity about the tax and social security position of those earning LETS;
- health and safety issues, including insurance and liability for accidents;

Box 8.3 continued

- consumer protection against poor-quality work;
- potential for illegal or immoral services to be offered through LETS.

References: Lang (1994); Letslink (2003).

Discussion point

1 To what extent does the unregulated and decentralised nature of this economic activity help, or hinder, LETS groups to meet the sustainability criteria: equity, futurity, valuing the environment?

supply and demand are predicated on the concept of the rational consumer – a creature who has perfect information about the range of goods on offer and is able to choose between them in a way that maximises his or her welfare. In globalised and postmodern markets, where brands compete as transient satisfiers of Maslow's higher-level needs, this assumption is questionable from an ecocentric perspective.

In short, ecological economics offers a holistic vision of a future sustainable economy, where human needs are met adequately by appropriate satisfiers in a way that respects the integrity of the natural world and the rights of future generations. It also offers mechanisms to achieve this, through community based initiatives appropriate to local contexts. The LETS concept is only one example of this – others include fair trading schemes to market developing world produce in the developed world; local credit unions to combat economic exclusion; worker and consumer co-operatives; farmers' markets to foster direct links between the producers and the consumers of food.

However, it all comes down to values in the end and the profound difficulty, in democratic states based on free-market principles, of changing the postmodern culture of the consumer society. The vision of ecological economics depends on consumers applying the sustainable development principles (equity, futurity, valuing the environment) in their economic decision decision making. Some argue that the 'Green consumerism' and 'ethical consumerism' phenomena are welcome and early signs of this. If people are willing to buy products advertised as 'fair traded' or 'environmentally friendly' this is an important first step towards a more sustainable pattern of consumption. The counter-argument, promulgated not least from the ecocentric point of view, dismisses these claims by categorising this as yet another example of branding, duping consumers into purchases, when what is needed is less consumption. Poor regulation means that many environmental claims by manufacturers are unfounded. In a postmodern world where consumption is a more significant means of expressing preferences than voting, Green consumerism may offer the affluent the prospect of spending for change, but this is denied to the poor, whose economic choices are constrained by bare necessity. Seen in these terms, Green consumerism becomes less a pathway to sustainability and more an irrelevant distraction from the real issues. Indeed, it can be argued that the most valuable contribution of ecological economics is its demonstration of the extent of change necessary to achieve a sustainable economy.

Further reading

Although the 1980s and 1990s saw the rapid development and application of economic ideas in environmental policy making, most mainstream economics textbooks pay little or no attention to the core issues considered in this chapter, except to introduce the basic theory of external costs and benefits. Scanning the index of such texts will yield an entry for 'externality' but not for 'sustainable development', 'environment' or even 'polluter pays principle'. However, to read in more detail than was possible here about micro-economic theory, Begg (2002) gives a good introduction, whilst Varian (1999: chapter 32) elaborates further for those already familiar with economic concepts such as Pareto efficiency. Ormerod (1994) offers a refreshing and readable antidote to the orthodoxy of standard economics textbooks. Pearce and Barbier (2000) is an accessible account of the recent growth of environmental economics and its potential role in achieving sustainable development. Roodman (1999) also provides an overview, using mainly American examples. Similar texts are Gilpin (2000) and Hussen (1999). Unpublished at the time of writing, Jordan and Wurzel (2003) promises an overview of novel environmental policy instruments, including environmental taxes and tradable permits, in seven industrialised countries. From the ecological economics perspective a useful introduction can be found in Faber *et al.* (1996). The classic text is Schumacher (1973), which sets out a very influential case against trends towards globalisation and for a different brand of economics.

9 Making policy for the environment – and for people

The analysis presented here of environmental problems and environmental policy has been unashamedly anthropocentric. The 6+ billions on the planet now will be increased by up to 5 billion more in the next fifty years. Ensuring even a basic minimum quality of life for the Earth's population will mean taking an instrumental approach to environmental capital and environmental services. There is nothing new in this – throughout history just such an approach has been taken. In the twenty-first century, however, an instrumental approach will necessarily involve the conservation of environmental capital in order to ensure the continued provision of environmental services, whether this is at the local level (Box 3.2), or at the global level, by implementing policies to limit the impacts of global warming (Box 2.2).

Environmental problems are human problems in two senses: humans define a problem once they become aware of a decline in environmental services; human behaviour is often the direct or indirect cause of that decline. Environmental policy offers a toolbox to fix (or at least begin to fix) environmental problems. This may include responses based on environmental management, but often the fundamental response is to get people to change their behaviour (Box 1.3).

There are tensions inherent in the relationship between people and the environment. The dexterity, language and cognitive abilities with which humans are endowed allow the overuse of environmental capital, but also give the capacity to imagine and plan for its future conservation. Human needs expressed in the context of postmodern consumer culture are driving unprecedented rates of resource extraction and waste production, but affluent lifestyles are linked with increased awareness of environmental problems.

Challenges for environmental policy makers are legion, but the toolbox can help. The precautionary principle and no-regrets strategies can provide the basis for preventative action when scientific information about an issue is limited. Appraisal of the appropriateness of technology can increase the chances that its use will create fewer problems than it solves. Life-cycle analysis, environmental management systems and sustainability management systems can assist in progress towards an economy which is less resource-intensive and therefore less wasteful. Government regulation has its place, but Green taxes and other economic instruments are potentially more effective (Box 8.2). More effective still are the techniques of persuasion (Boxes 3.1, 8.3), although persuasion can become coercion (Box 6.3).

The impossibility of halting economic growth in a world of poverty and continuing population growth has led to the aspirational concept of sustainable development. The principles of sustainable development – equity, futurity, valuing the environment – can be stated very simply, but in the fifteen years since their almost universal adoption at the Rio Summit progress has been limited, more or less, to the development of initiatives and action plans. Barriers to economic growth in less developed countries include indebtedness and the rules which govern international trade. These barriers preclude sustainable economic development and sustainable environmental management (Box 7.2).

In the developed world environmental quality is increasing as local and regional issues are addressed (Box 7.1). The Montreal Protocol to control ozone-depleting substances (Box 5.2), and the Kyoto Protocol on greenhouse gas emissions (Box 2.2), offer rare examples of implementation being followed through at the global level. For the Kyoto Protocol, this success can be counted as only partial since the withdrawal of the United States and Australia from the agreement.

Vested interest and short-termism are probably the two most significant barriers to the implementation of policies for sustainable development (Boxes 2.1–2, 3.1, 4.1, 5.1). The effective formulation and implementation of environmental policies will require leadership if barriers are to be overcome and fundamental changes achieved. Not just leadership by government, although that will be essential. But leadership also within policy sectors, interest groups, communities, corporations and every other sort of network where people have the possibility of influencing others by the example they set. Remembering always that there are two alternatives to sustainable development: unsustainable development or no development at all.

References

Abraham, John (1991) *Food and Development*, London: Kogan Page/WWF.
AccountAbility, Forum for the Future and British Standards Institute (2001) *The Sigma Project: Sustainability in Practice*, <http://www.projectsigma.com>.
Adams, W. M. (2001) *Green Development*, London: Routledge.
Andersen, M. S. and Massa, I. (2000) 'Ecological modernisation: origins, dilemmas and future directions', *Journal of Environmental Policy and Planning*, 2: 337–45.
Beck, U. (1992) *Risk Society*, London: Sage.
Begg D. (2002) *Economics*, London: McGraw-Hill (7th edn).
Benson, J. (2000) *Environmental Ethics*, London: Routledge.
Bird E. C. F. (1993) *Submerging Coasts: The Effects of a Rising Sea Level on Coastal Environments*, Chichester: Wiley.
Blowers, A., Lowry, D. and Solomon, B. D. (1991) *The International Politics of Nuclear Waste*, London: Macmillan.
Boehmer-Christiansen, S. (2001) 'Global warming: science as a legitimator of politics and trade?', in Handmer, J. W., Norton, T. W. and Dovers, S. R., *Ecology, Uncertainty and Policy*, Harlow: Prentice Hall 116–37.
Boehmer-Christiansen, Sonja and Skea, Jim (1991) *Acid Politics: Environmental and Energy Politics in Britain and Germany*, London: Belhaven Press.
Bookchin, M. (1974) *Post-scarcity Anarchism*, London: Wildwood House.
Bookchin, M. (1990) *The Philosophy of Social Ecology*, Montreal: Black Rose Books.
BP (2002) *Statistical Review of World Energy*, London: BP, <http://www.bp.com/centres/energy/>.
Braverman, Harry (1974) *Labor and Monopoly Capital: The Degradation of Work in the Twentieth Century*, London: Monthly Review Press.
Braybrooke, D. and Lindblom, C. (1963) *A Strategy of Decision: Policy Evaluation as a Social Process*, New York: Free Press.
Brimblecombe, P. (1987) *The Big Smoke: A History of Air Pollution in London since Medieval Times*, London: Routledge.
British Standards Institute (1997) *The EMS Handbook: A Guide to the BS EN ISO 14000 series*, 2nd edn, London: BSI.
British Standards Institute (2001) *The Sigma Project*, <http://www.sigmaproject.com>.

Bronowski, J. (1973) *The Ascent of Man*, London: BBC.

Brown, L. R. (2001) *Eco-economy: Building an Economy for the Earth*, London: Earthscan.

CAG Consultants (1993) *A Guide to the Eco-management and Audit Scheme for UK Local Government*, London: HMSO.

CAG Consultants (1997) *Environmental Capital: A New Approach*, Cheltenham: Countryside Commission.

Carson, R. (1962) *Silent Spring*, Boston, MA: Houghton Mifflin.

Cawson, A. and Saunders, P. (1983) 'Corporatism, competitive politics and class struggle', in King, R. (ed.) *Capital and Politics*, London: Routldge.

CEROI (2001) *Core Indicators*, <http://www.ceroi.net/ind/matrix.asp>.

Chapple, C. K. and Tucker, M. E. (eds) (2000) *Hinduism and Ecology: The Intersection of Earth, Sky and Water*, Cambridge MA: Harvard University Press.

Chase, A. (1980) *The Legacy of Malthus: The Social Cost of the New Scientific Racism*, Urbana IL: University of Illinois Press.

Chesters, G. and Konrad, J. F. (1971) 'Effects of pesticide usage on water quality', *Bioscience*, 21: 565–9.

Chudleigh, R. and Cannell, W. (1984) *Radioactive Waste: The Gravedigger's Dilemma*, London: Friends of the Earth.

Connelly, J. and Smith, G. (1999) *Politics and the Environment: From Theory to Practice*, London: Routledge.

Conway, R. and Pretty, Jules, N. (1991) *Unwelcome Harvest: Agriculture and Pollution*, London: Earthscan.

Corbridge, Stuart (1993) *Debt and Development*, Oxford: Blackwell.

Costanza, R. (1989) 'What is ecological economics?', *Ecological Economics*, 1: 1–7.

Cotgrove, Stephen and Duff, Andrew (1981) 'Environmentalism, values and social change', *British Journal of Sociology*, 32: 92–110.

Countryside Agency (2002) <http://www.qualityoflifecapital.org.uk>.

Crenson M. (1971) *The Unpolitics of Air Pollution*, Baltimore MD: Johns Hopkins University Press.

Cunningham, G. (1963) 'Policy and practice', *Public Administration*, 41: 229–37.

Dahl, R. (1961) *Who Governs?*, New Haven: Yale University Press.

Daly, M. (1987) *Gyn/Ecology*, London: The Women's Press.

Dawkins, R. (1976) *The Selfish Gene*, Oxford: Oxford University Press.

Deffeyes, K. S. (2001) *Hubbert's Peak: The Impending World Oil Shortage*, Princeton NJ: Princeton University Press.

Department for Environment, Food and Rural Affairs (2000) *The Air Quality Strategy for England, Scotland, Wales and Northern Ireland*, Cm 4548, London: Stationery Office, <http://www.defra.gov.uk/environment/airquality/strategy/index.htm>.

Department for International Development (1997) *Eliminating World Poverty: A Challenge for the Twenty-first Century*, London: DFID.

Department for Trade and Industry (2003) *Our Energy Future: Creating a Low Carbon Economy*, London: DTI.

Department of Energy (1989) *The Severn Barrage Project: General Report* (Energy Paper 57), London: HMSO.

Department of the Environment (1993) *Externalities from Landfill and Incineration* London: HMSO.

Desai, V. and Potter, R. B. (eds) (2002) *The Companion to Development Studies*, London: Arnold.

de-Shalit, A. (1995) *Why Posterity Matters*, London: Routledge.

Dobson, A. (ed.) (1999) *Fairness and futurity*, Oxford: Oxford University Press.

Douthwaite, B. (2002) *Enabling Innovation: A Practical Guide to Understanding and Fostering Technological Change*, London: Zed Books.

Dovers, S. R., Norton, T. W. and Handmer, J. W. (2001) 'Ignorance, uncertainty and ecology: key themes', in Handmer, J. W., Norton, T. W. and Dovers, S.R., *Ecology, Uncertainty and Policy*, Harlow: Prentice Hall 1–25.

Downs, Anthony (1972) 'Up and down with ecology: the issue-attention cycle', *Public Interest*, 28: 38–50.

Dror, Y. (1964) 'Muddling through: "science" or inertia?', *Public Administration Review*, 24: 153–7.

Easton D. (1965) *A Systems Analysis of Political Life*, New York: Wiley.

Ecologist, The (1993) *Whose Common Future?* London: Earthscan.

ECOTEC Research and Consulting (1998) *Effectiveness of the Landfill Tax in the UK*, <http://www.foe.co.uk/resource/reports/effectiveness_landfill_tax.pdf>.

EEC (1993) Council Regulation No. 1836/93, 'Community Eco-Management and Audit Scheme', *Official Journal of the European Communities* No. L168, 10 July.

Elkington, J. (2001a) 'The "triple-bottom line" for twenty-first century business', in Starkey, R. and Welford, R. (eds) *The Earthscan Reader in Business and Sustainable Development*, London: Earthscan 20–43.

Elkington, J. (2001b) *The Chrysalis Economy*, London: SustainAbility.

Elkington, J. and Hailes, J. (1988) *The Green Consumer Guide*, London: Gollancz.

Environmental Data Services (2001) 'Secondary aggregates: gearing up for new market opportunities', *ENDS Report* 323: 29–31.

Erickson, S. (1999) *Fundamentals of Environmental Management*, Chichester: Wiley.

Etzioni, A. (1967) 'Mixed-scanning: a "third" approach to decision making', *Public Administration Review*, 27: 385–92.

Faber, M., Manstetten, R. and Proops, J. (1996) *Ecological Economics: Concepts and Methods*, Cheltenham: Edward Elgar.

Factor Ten Club (1994) *The Carnoules Declaration*, Wuppertal: Factor Ten Club.

FCCC Secretariat (1992) *Framework Convention on Climate Change*, Bonn: United Nations Framework Convention on Climate Change.

FCCC Secretariat (1997) *FCCC/CP/1997/CRP.4 (9 December)*, Bonn: FCCC Secretariat.

Ferguson, N. M., Ghani, A. C., Donnelly, C. A., Hagenaars, T. J. and Anderson, R. M. (2002) 'Estimating the human health risk from possible BSE infection of the British sheep flock', DOI: 10.1038/nature709, 9 January.

Fernie, John and Pitkethly, Al (1985) *Resources: Environment and Policy*, London: Harper & Row.

Finnish Forest Industries Federation (2001) <http://www.forestindustries.fi/>.

Finnish Nature League (2002) <http://www.luontoliitto.fi/forest/>.

Francioni, F. (ed.) (2001) *Environment, Human Rights and International Trade*, Oxford: Hart.

Galbraith, J. K. (1987) *The Affluent Society*, 4th edn, London: Pelican.

Gandy, Matthew (1994) *Recycling and the Politics of Urban Waste*, London: Earthscan.

Gaston, K. J. (ed.) (1996) *Biodiversity*, Oxford: Blackwell.

Gibson, C., McKean, M. and Ostrom, E. (2000) *People and Forests: Communities, Institutions and Governance*, Cambridge MA: MIT Press.

Giddens, Anthony (1999) *Runaway World: How Globalisation is Reshaping our Lives*, London: Profile.

Gilpin, A. (2000) *Environmental Economics*, Chichester: Wiley.

Gleckman, Harris and Krut, Riva (1997) 'Neither international nor standard: the limits of ISO 14001 as an instrument of global corporate environmental management', in C. Sheldon (ed.) *ISO 14001 and beyond*, Sheffield: Greenleaf.

Goodin, R. (1982) 'Discounting discounting', *Journal of Public Policy*, 2: 53–72.

Goudie, A. (2000) *The Human Impact on the Natural Environment*, 5th edn, Oxford: Blackwell.

Gouldson, A. and Murphy, J. (1997) 'Ecological modernisation: restructuring industrial economies', in Jacobs, M. (ed.) *Greening the Millennium?* Oxford: Blackwell 74–86.

Gray, T. (1995) (ed.) *UK Environmental Policy in the 1990s*, London: Macmillan.

Green, G. M. and Sussman, R. W. (1990) 'Deforestation history of the eastern rain forests of Madagascar from satellite images', *Science*, 248: 212–15.

Grove-White, R. (1993) 'Environmentalism: a new moral discourse?', in Milton, K. (ed.) *Environmentalism: The View from Anthropology*, London: Routledge 18–30.

Ham, C. and Hill, M. (1993) *The Policy Process in the Modern Capitalist State*, Hemel Hempsted: Wheatsheaf.

Hardee-Cleaveland, K. and Banister, J. (1988) 'Fertility policy and implementation in China, 1986–88', *Population and Development Review*, 14: 245–86.

Hardin, G. (1968) 'The tragedy of the commons', *Science*, 162: 1243–8.

Hardin, G. (1974) 'Living on a lifeboat', *BioScience*, 24 (10): 561–8.

Harremoës, P., Gee, D., MacGarvin, M., Stirling, A., Keys, J., Wynne, B. and Guedes-Vaz, S. (eds) *The Precautionary Principle in the Twentieth Century*, London: Earthscan.

Harrison, R. (ed.) (2001) *Pollution: Causes, Effects and Control*, Cambridge: Royal Society of Chemistry.

Hart, S. L. (1997) 'Beyond greening: strategies for a sustainable world', in

Harvard Business Review on Corporate Strategy Boston MA: Harvard
Business School 121–46.

Hartmann, B. (1995) *Reproductive Rights and Wrongs*, Boston MA: South End
Press.

Hawke, N. (2002) *Environmental Policy: Implementation and Enforcement*,
Aldershot: Ashgate.

Hawken, P., Lovins, A. and Lovins, H. (1999) *Natural Capitalism*, Snowmass
CO: Rocky Mountain Institute.

Heilbroner, R. (1977) *Business Civilisation in Decline*, Harmondsorth: Penguin.

Henderson, Caspar (1998) 'Take your partners for the Natural Step', *Green
Futures*, November/December: 38–40.

Hitchcock, D. and Blair, A. (2000) *Environment and Business*, London:
Routledge.

Hoogvelt, A. (1997) *Globalisation and the Postcolonial World: The New
Political Economy of Development*, Basingstoke: Macmillan.

Hopfenbeck, W. (1993) *The Green Management Revolution*, London: Prentice
Hall.

Hortensius, D. and Barthel, M. (1997) 'Beyond 14001: an introduction to the ISO
14001 series', in C. Sheldon *ISO 14001 and beyond*, Sheffield: Greenleaf
19–44.

Houghton, J., Jenkins, G., and Ephraums, J. (eds) (1990) *Scientific Assessment of
Climate change: Report of Working Group I*, Cambridge: Cambridge
University Press.

Houghton, J., Ding, Y., Griggs, D., Noguer, M., van der Linden P. and Xiaosu,
D. (eds) (2001a) *Climate Change 2001: The Scientific Basis. Contribution of
Working Group I to the Third Assessment Report of the Intergovernmental
Panel on Climate Change (IPCC), Technical Summary*, Geneva: IPCC or
<http://www/ipcc.org.ch>.

Houghton, J., Ding, Y., Griggs, D., Noguer, M., van der Linden P. and Xiaosu,
D. (eds) (2001b) *Climate Change 2001: The Scientific Basis. Contribution of
Working Group I to the Third Assessment Report of the Intergovernmental
Panel on Climate Change (IPCC), Summary for policy makers*, Geneva: IPCC
or <http://www/ipcc.org.ch>.

Howes, Chris (1997) *The Spice of Life: Biodiversity and the Extinction Crisis*,
London: Blandford Press.

Howes, R., Skea, J. and Whelan, B. (1997) *Clean and Competitive? Motivating
Environmental Performance in Industry*, London: Earthscan.

Hussen, A. M. (1999) *Principles of Environmental Economics*, London:
Routledge.

Ikein, Augustine (1990) *The Impact of Oil on a Developing Country: The Case of
Nigeria*, London: Praeger.

Illich, I. (1973) *Tools for Conviviality*, London: Fontana.

Inglehart, Ronald (1977) *The Silent Revolution*, Princeton NJ: Princeton
University Press.

Irwin, A. (1995) *Citizen Science*, London: Routledge.

Jackson, T. (1996) *Material Concerns*, London: Routledge.

Jacobs, Jane M. (1993) '"Shake 'im this country": the mapping of the Aboriginal sacred in Australia – the case of Coronation Hill', in Jackson, Peter and Penrose, Jan, *Constructions of Race, Place and Nation*, London: UCL Press 100–18.

Jacobs, M. (1991) *The Green Economy: Environment, Sustainable Development and the Politics of the Future*, London: Pluto.

Jenkins, W. (1978) *Policy Analysis*, Oxford: Martin Robertson.

Jeremy, D. J. (1995) 'Corporate responses to the emergent recognition of a health hazard in the UK asbestos industry: the case of T&N, 1920–1960', *Business and Economic History*, 24: 254–65.

Jones, B. (1998) 'Pressure groups', in Jones, B. (ed.) *Politics UK*, Hemel Hempstead: Prentice Hall.

Jordan, A. and O'Riordan, T. (2000) 'Environmental politics and policy processes', in O'Riordan, T. (ed.) *Environmental Science for Environmental Management*, Harlow: Prentice Hall.

Jordan, A. and Richardson, J. (1987) *British Politics and the Policy Process*, London: Unwin Hyman.

Jordan, A., Wurzel, R. and Zito, A. (2003) *New Instruments of Environmental Governance?* special issue of the journal *Environmental Politics* 12 (1).

Kahn, H., Brown, W., and Martel, L. (1976) *The next 200 years: Scenario for America and the World*, New York: Morrow.

Klein, N. (1999) *No Logo*, London: Flamingo.

Korten, D. (1995) *When Corporations rule the World*, West Hartford CT: Kumarian Press.

Kramer, R. A., Richter, D. D., Pattanayak, S. and Sharma, N. P. (1997) 'Ecological and economic analysis of watershed protection in eastern Madagascar', *Journal of Environmental Management*, 49: 277–95.

Kunin, W. E. and Lawton, J.H. (1996) 'Does biodiversity matter? Evaluating the case for conserving species', in Gaston, K. J. (ed.) *Biodiversity*, Oxford: Blackwell 283–308.

Lamb, R. (1996) *Promising the Earth*, London: Routledge.

Lang, Peter (1994) *Let's Work: Rebuilding the Local Economy*, Bristol: Grover Books.

Leisinger, K. M. (1994) *All our People: Population Policy with a Human Face*, Washington DC: Island Press.

Leisinger, K., Schmitt, K. and Pandya-Lorch, R. (2002) *Six Billion and Counting: Population Growth and Food Security in the Twenty-first Century*, Washington DC: International Food Policy Research Institute.

Letslink (2003) <http://www.letslinkuk.org>.

Lomborg, B. (2001) *The Skeptical Environmentalist*, Cambridge: Cambridge University Press.

Lovelock, J. (1991) *Gaia: The Practical Science of Planetary Medicine*, London: Gaia Books.

Lowe, Philip and Goyder, Jane (1983) *Environmental Groups in Politics*, London: Allen & Unwin.

Lukes, S. (1974) *Power: A Radical View*, London: Macmillan, chapter 7.

Lyon, D. (1999) *Postmodernity*, 2nd edn, Buckingham: Open University Press.

Lyotard, J.-F. (1984) *The Postmodern Condition*, Manchester: Manchester University Press.

MacGarvin, M. (2002) 'Fisheries: taking stock', in Harremoës, P., Gee, D., MacGarvin, M. Stirling, A., Keys, J., Wynne, B. and Guedes-Vaz, S. (eds) *The Precautionary Principle in the Twentieth Century*, London: Earthscan 10–25.

MacGillivray, A. and Walker, P. (2000) 'Local social capital: making it work on the ground', in Baron, S., Field, J. and Schuller, T. (eds) *Social Capital: Critical Perspectives*, Oxford: Oxford University Press.

Maharaj, Niala and Dorren, Gaston (1995) *The Game of the Rose: The Third World in the Global Flower Trade*, Utrecht, Netherlands: International Books for the Institute for Development Research, Amsterdam.

Malthus, T. (1803) *An Essay on the Principles of Population*, selected and introduced by Winch, D. (1992) Cambridge: Cambridge University Press.

Marsh, D. and Rhodes, R. (1992) *Policy Networks in British Government*, Oxford: Oxford University Press.

Martell, L. (1994) *Ecology and Society: An Introduction*, Cambridge: Polity Press.

Maslow, Abraham (1970) *Motivation and Personality*, New York: Harper & Row.

Max-Neef, M. A. (1991) *Human Scale Development*, London: Apex Press.

Mazur, L. (ed.) (1994) *Beyond the Numbers: A Reader on Population, Consumption and the Environment*, Washington DC: Island Press.

McCormick, J. (1995) *The Global Environmental Movement*, Chichester: Wiley.

McKean, Margaret A. (2000) 'Common property: what is it, what is it good for, and what makes it work?', in Gibson, C., McKean, M. and Ostrom, E. (eds) *People and Forests: Communities, Institutions and Governance*, Cambridge MA: MIT Press 27–55.

McLaren, D., Bullock, S., and Yousuf, N. (1998) *Tomorrow's World*, London: Earthscan.

Meadows, Donella H. *et al.* (1972) *The Limits to Growth: A Report for the Club of Rome's Project on the Predicament of Mankind*, London: Earth Island.

Meadows, D. H., Meadows, D. L. and Randers, J. (1992) *Beyond the Limits*, London: Earthscan.

Miliband, R. (1969) *The State in Capitalist Society*, London: Weidenfeld & Nicolson.

Millar, C. (2001) 'Green funds and their growth and influence on corporate environmental strategy', in Hillary, R. (ed.) *Environmental Management Handbook: The Challenge for Business*, London: Earthscan 49–52.

Miller, G. T. (2001) *Living in the Environment*, Pacific Grove CA: Brooks Cole.

Mitchell, C. (2000) 'The England and Wales non-fossil fuel obligation: history and lessons', *Annual Review of Energy and the Environment*, 25: 285–312.

Mol, A. and Sonnenfeld, D. (2000) *Ecological Modernisation around the World*, special issue of the journal *Environmental Politics* 9 (1).

Murphy, D. F. and Bendell, J. (2001) 'Getting engaged: business–NGO relations on sustainable development', in Starkey, R. and Welford, R. (eds) *The Earthscan Reader in Business and Sustainable Development*, London: Earthscan 288–312.

Myers, N. (ed.) (1994) *The Gaia Atlas of Planet Management*, London: Gaia Books.

Naess, A. (1973) 'The shallow and the deep, long range ecology movement: a summary', *Inquiry*, 16: 95–100.

Naess, A. (1988) 'The basics of deep ecology', *Resurgence*, 126: 4–7.

National Society for Clean Air (2002) *Pollution Handbook*, Brighton: NSCA.

Nelson, Jane, Zollinger, Peter and Singh, Alok (2001) *The Power to Change*, London: International Business Leaders' Forum and SustainAbility.

Newhouse, M. L. and Thompson H. (1965) 'Mesothelioma of pleura and peritoneum following exposure to asbestos in the London area', *British Journal of Industrial Medicine*, 22: 261–9.

North, R. D. (1995) *Life on a Modern Planet*, Manchester: Manchester University Press.

Oak Ridge National Laboratory (2002) *Simulated Eruption of Mt Vesuvius*, <http://www.ornl.gov/gist/projects/LandScan/STATE/Index.htm>.

Office of Civilian Radioactive Waste Management (2003) *Yucca Mountain Project*, <http://www.ocrwm.doe.gov/ymp/index.shtml>.

O'Neill, J. (1993) *Ecology, Policy and Politics: Human Wellbeing and the Natural World*, London: Routledge.

Ophuls, W. (1977) *Ecology and the Politics of Scarcity*, San Francisco: Freeman.

O'Riordan, T. (1995) 'Core beliefs and the environment', *Environment*, 37 (8): 4–9, 25–9.

Ormerod, P. (1994) *The Death of Economics*, London: Faber.

Parfit, D. (1983) 'Energy policy and the further future: the social discount rate', in Maclean, D. and Brown, P. (eds) *Energy and the Future*, Totowa NJ: Rowman & Littlefield.

Park, C. (2001) *The Environment: Principles and Applications*, London: Routledge.

Pearce, D. and Barbier, E. (2000) *Blueprint for a Sustainable Economy*, London: Earthscan.

Pearce, D., Markandya, A. and Barbier, E. (1989) *Blueprint for a Green Economy*, London: Earthscan.

Pearce, F. (1991) *Green Warriors*, London: Bodley Head.

Pierce-Colfer, Carol J. and Resosudarmo, Pradnja (eds) (2001) *Which Way Forward? People, Forests, and Policymaking in Indonesia*, Washington DC: RFF Press.

Pepper, David (1996) *Modern Environmentalism*, London: Routledge.

Pepper, I., Gerba, C. and Brusseau, M. (1996) *Pollution science*, San Diego CA: Academic Press.

Pitman, Nigel C. A. and Jørgensen, Peter M. (2002) 'Estimating the size of the world's threatened flora', *Science*, 1 November: 2002: 989.

Ponting, Clive (1991) *A Green History of the World*, London: Penguin Books.

Popper, K. (1965) *The Logic of Scientific Discovery*, New York: Harper & Row.

Porritt J. (1997) 'Environmental politics: the old and the new', in Jacobs M. (ed.) *Greening the Millennium? The New Politics of the Environment*, Oxford: Political Quarterly 62–73.

Porritt, J. (2000) *Playing Safe: Science and the Environment*, London: Thames & Hudson 62–73.

Powell, J. (1998) *Postmodernism for Beginners*, London: Writers & Readers.

Radcliffe, J. (2002) *Green Politics*, Basingstoke: Palgrave.

Rawcliffe, P. (1998) *Environmental Pressure Groups in Transition*, Manchester: Manchester University Press.

Reaka-Kudla, M. L., Wilson, D. E. and Wilson, E. O. (eds) (1997) *Biodiversity II*, Washington DC: Joseph Henry Press.

Redclift, M. (1987) *Sustainable Development: Exploring the Contradictions*, London: Methuen.

Redclift, M. (1996) *Wasted: Counting the Costs of Global Consumption*, London: Earthscan.

Reinhardt, F. (1992) 'Du Pont Freon® Products Division', in Buchholz, R., Marcus, A. and Post, J. (eds) *Managing Environmental Issues: A Casebook*, Englewood Cliffs NJ: Prentice Hall 261–85.

Rifkin, Jeremy (1997) *The Biotech Century: How Genetic Commerce will Change the World*, London: Phoenix.

Roberts, J. (1991) 'Clarity, ambivalence or confusion? An assessment of pressure group motives at the Hinkley Point "C" Public Inquiry', *Energy and Environment*, 2: 45–67.

Roberts, J., Elliott, D. and Houghton, T. (1991) *Privatising Electricity*, London: Belhaven Press.

Roberts, N. (1998) *The Holocene: An Environmental History*, Oxford: Blackwell.

Roodman D. (1999) *The Natural Wealth of Nations*, London: Earthscan.

Rose, S. (1984) *Not in Our Genes*, Harmondsworth: Penguin.

Royal Commission on Environmental Pollution (1985) *Managing Waste: The Duty of Care*, Cmnd 9675, London: HMSO.

Sabatier P. A. (1986) 'Top-down and bottom-up approaches to implementation research: a critical analysis and suggested synthesis', *Journal of Public Policy*, 6: 21–48.

Schmitter, P. (1979) 'Still the century of corporatism?', in Schmitter, P. and Lembruch, G., *Trends Towards Corporatist Intermediation*, London: Sage 7–52.

Schumacher, E. F. (1973) *Small is Beautiful*, London: Vintage.

Schuurman, F. J. (ed.) (2001) *Globalization and Development Studies*, London: Sage.

Selman, P. (1996) *Local Sustainability*, London: Paul Chapman.

Sheldon, C. (ed.) (1997) *ISO 14001 and beyond*, Sheffield: Greenleaf.

Shennan, I. (1992) 'Impacts of sea-level rise on the Wash, United Kingdom', in Tooley, M. J. and Jelgersma, S. (eds) *Impacts of Sea-level Rise on European Coastal Lowlands*, Oxford: Blackwell 72–93.

Shiva V. (1994) 'Development, ecology and women', in Merchant C. (ed.)

Ecology: Key Concepts in Critical Theory, Atlantic Highlands NJ: Humanities Press 272–80.

Simon, H. (1945) *Administrative Behaviour*, Glencoe IL: Free Press.

Simon, J. (1997) *The Ultimate Resource 2*, Princeton NJ: Princeton University Press.

Singer, P. (1983) *Animal Liberation: Towards an End to Man's Inhumanity to Animals*, Wellingborough: Thorson.

Singer, P. (2002) *One World: The Ethics of Globalization*, New Haven CT: Yale University Press.

Stiglitz, J. (2002) *Globalisation and its Discontents*, London: Allen Lane.

Stork, N. E. (1997) 'Measuring global biodiversity and its decline', in M. L. Reaka-Kudla *et al.* (eds) *Biodiversity II*, Washington DC: Joseph Henry Press 41–68.

Strategy Unit (2002) *Waste Not, Want Not: A Strategy for Tackling the Waste Problem in England*, London: Cabinet Office.

Sutton, P. (2000) *Explaining Environmentalism*, Aldershot: Ashgate.

Tata Research Institute (2002) <http://www.teriin.org/climate/nego.htm>.

The Natural Step (2002) <http://www.naturalstep.org/framework/systemconditions.pdf >.

Toulmin, Camilla, Scorer, Ian and Bishop, Joshua (1995) 'The future of Africa's drylands', in Kirkby, John, O'Keefe, Phil and Timberlake, Lloyd, *The Earthscan Reader in Sustainable Development*, London: Earthscan 254–5.

Turner, R. K. (1993) 'Sustainability: principles and practice', in Turner, R. K. (ed.) *Sustainable Environmental Economics and Management*, London: Belhaven Press 3–36.

Tweedale, G. (2000) *Magic Mineral to Killer Dust: T&N and the Asbestos Hazard*, Oxford: Oxford University Press.

UN Conference on Environment and Development (1993) *Agenda 21: The United Nations Programme of Action from Rio*, New York: United Nations.

UN Economic Commission for Europe (2000) *2000 Review of Strategies and Policies for Air Pollution Abatement*, Geneva: UNECE.

UN Environment Programme (2000a) *Action on Ozone*, Nairobi: UNEP.

UN Environment Programme (2000b) *GEO-2000 Global Environment Outlook*, <http://www.grid.unep.ch/geo2000/english/0149.htm>.

UN Environment Programme (2002) <www.unep.ch/etu/doha/papers.htm>.

UNFPA (1999) *The State of the World Population, 1999: 6 Billion: A Time for Choices*, New York: UNFPA.

UNFPA (2001) *The State of the World Population, 2001: Footprints and Milestones: Population and Environmental Change*, New York: UNFPA.

UNFPA (2002) personal communication 7 March.

US Environmental Protection Agency (1993) *The Plain English Guide to the Clean Air Act*, Research Triangle Park NC: USEPA, <http://www.epa.gov/oar/oaqps/peg_caa/pegcaain.html>.

Utting, P. (2002) *The Greening of Business in Developing Countries*, London: Zed Books.

Valleron, Alain-Jacques, Boelle, Pierre-Yves, Will, Robert and Cesbron, Jean-Yves (2001) 'Estimation of epidemic size and incubation time based on age characteristics of vCJD in the United Kingdom', *Science*, 294: 1726–8.

Varian, H. R. (1999) (5th edn) *Intermediate Microeconomics: A Modern Approach*, London: Norton.

Vogler, J. (2000) *The Global Commons*, Chichester: Wiley.

von Weizsäcker, E., Lovins, A. and Lovins, H. (1997) *Factor Four: Doubling Wealth, Halving Resource Use*, London: Earthscan.

Wackernagel, M. and Rees, W. (1996) *Our Ecological Footprint*, Gabriola Island: New Society.

Wilson, Edward O. (1975) *Sociobiology: The New Synthesis*, Cambridge MA: Harvard University Press.

Wagner, L. A., Sullivan, D. E. and Sznopek, J. L. (2002) *Economic Drivers of Mineral Supply*, USGS <http://pubs.usgs.gov/of/2002/of02-335/>.

Wilson, Edward O. (1992) *The Diversity of Life*, London: Penguin Books.

Wilson, Edward O. (2002) *The Future of Life*, London: Little Brown.

Winch, D. (1992) 'Introduction', to Malthus, T., *An Essay on the Principles of Population*, Cambridge: Cambridge University Press.

Winter, M. (1996) 'Intersecting departmental responsibilities, administrative confusion and the role of science in government: the case of BSE', *Parliamentary Affairs*, 49: 550–65.

Wolf, Arthur (1986) 'The pre-eminent role of government intervention in China's family revolution', *Population and Development Review*, 12: 101–16.

Wolfe, A. (1977) *The Limits of Legitimacy*, New York: Free Press.

World Bank (2002) *World Development Indicators, 2002*, Washington DC: World Bank, <http://www.worldbank.com/data/wdi2002/ >.

World Commission on Environment and Development (1987) *Our Common Future*, Oxford: Oxford University Press.

World Conservation Monitoring Centre (1992) *Global Biodiversity*, London: Chapman & Hall.

World Development Movement (2001) *The Tricks of the Trade: How Trade Rules Are Loaded against the Poor*, <http://www.wdm.org.uk/cambriefs/Wto/tricks.pdf>.

World Health Organisation (1971) *Solid Waste Disposal and Control*, Geneva: WHO.

World Summit on Sustainable Development (2002) *The Johannesburg Summit Test: What will Change?* <http://www.johannesburgsummit.org/index.html>.

World Trade Organisation (1999) *The World Trade Organisation in Brief*, Geneva: WTO <http://www.wto.org/>.

World Trade Organisation (2001) *Trading into the Future*, Geneva: WTO, <http://www.wto.org/>.

Xerox Corporation (1998) *Business and the Environment: Solutions for a Changing World*, New York: Xerox Corporation.

Young, Elspeth (1995) *Third World in the First*, London: Routledge.

Index

Aboriginal peoples 56–7
accommodators 59–60
acid emissions 22, 109, 161, 168, 169–70
acid rain 141
advertising 48, 62
afforestation 52
Africa, urbanisation 30
Agenda 21 action plan 2, 184–5
agriculture: early human societies 44; Green revolution 103–4; high-yielding varieties 103–4; improvements in 27; land resources 8; land use patterns 32; Madagascar 80; Malthusianism 67; subsidies 181, 204; technology 101, 103–4, 105; *see also* pesticides
Air BP 136–7
air pollution 18, 22, 24; trans-boundary 168, 169–70; US regulation 149; waste concentration 30
anthropocentrism 37, 59, 62, 217
appropriate technology 104–6, 117, 218
asbestos 121–2
assimilation of waste 8, 21, 23–4, 85, 106, 113
asthma 24
auditing, environmental 128
Australia: Coronation Hill case study 56–7; Kyoto Protocol rejection 53, 185, 218; LETS schemes 214; The Natural Step 135

backcasting 136
Barthel, M. 130, 131
BAT *see* Best Available Techniques
BATNEEC *see* Best Available Techniques Not Entailing Excessive Costs
belief systems 3
benchmarks 137

Best Available Techniques (BAT) 86, 87–8, 157
Best Available Techniques Not Entailing Excessive Costs (BATNEEC) 86–7, 88, 132, 133, 137
Best Practicable Environmental Option (BPEO) 4, 86, 87, 88, 111–13, 115
Best Practical Means approach 86
bio-accumulation 23, *23*
bio-degradation 23–4
biodiversity 7, 30–2, 186, 210; forests 78, 187; Global Environment Facility 185; wetlands 81
Biodiversity Convention 186, 187
Bird, E. C. F. 25
birth control 28, 69, 70, 162, 163
bovine spongiform encephalopathy (BSE) 92, 94, 95–6, 100–1
BPEO *see* Best Practicable Environmental Option
Brent Spar 126
British Standards Institute (BSI) 131
Brundtland Commission 73–4, 165, 184
Brundtland, Gro Harlem 73, 75
BSE *see* bovine spongiform encephalopathy
BSI *see* British Standards Institute

CAMPFIRE project 62, 76–7
Canada, LETS schemes 214
capital *see* environmental capital; social capital
capitalism 55, 118, 148, 173, 188; *see also* free trade; market economics
carbon cycle 21–2
carbon dioxide *16, 17,* 21–2; environmental space analysis *85*; Framework Convention on Climate Change 51–3; nuclear power 106; organic waste 24; sea-level rise 27; Severn barrage 106, 109; United States/India comparison 28; *see also*

global warming; greenhouse gases
Carson, Rachel 20
cause groups 144, 165, 166
CBA see cost–benefit analysis
CERES see Coalition for
    Environmentally Responsible
    Economies
CFCs (chlorofluorocarbons) 122–5, *124*
China, population control 161, 162–3
chlorofluorocarbons (CFCs) 122–5, *124*
CJD see Creutzfeld Jakob disease
climate: Framework Convention on
    Climate Change 51–3, 185; Global
    Environment Facility 185;
    Intergovernmental Panel on Climate
    Change 22, 63, 96–7; see also global
    warming
Co-operative Bank 136
coal 7, 10, *11*, *13*, *16*, 154, 161
Coalition for Environmentally
    Responsible Economies (CERES)
    133
coastal erosion 25–7
colonialism 62
combustion of waste 24
common property regimes 54, *55*
Communal Area Management
    Programme for Indigenous Resource
    Exploitation (CAMPFIRE) project
    62, 76–7
community 213, 214, 215
competition, economic 190, 191
Connelly, J. 145
conservation 58, 59
constant environmental capital 83, 84,
    211
consumer culture 48–9, 55, 215, 216,
    217; see also consumption
consumers: Green consumerism 125,
    215, 216; micro-economic theory
    189–90, 191–3; pressure from 119,
    123, 125–6; product charges 202;
    revealed preferences 206; service
    approach 115–16; stated preferences
    207
consumption: advertising 62; economic
    efficiency 191; environmental space
    analysis 85, *85*, 86; postmodern
    society 47, 48–9, 59, 63, 215–16;
    satisfiers 45, 48–9; technology impact
    on 101, 110; see also consumer
    culture
contingent valuation studies 207
contraception 28, 162, 163; see also birth
    control

Convention on Long-range
    Trans-boundary Air Pollution
    169–70
copper *10*, *13*, *17*
cornucopians 60, 68
Coronation Hill 56–7
corporate environmental policy making
    118–38
corporate responsibility 118, 125
corporatism 148, 165
cost–benefit analysis (CBA) 207–8
cost-effectiveness 2, 99
Cotgrove, Stephen 61, 64
Council for the Protection of Rural
    England (CPRE) 145
Cox, Vivienne 136–7
CPRE see Council for the Protection of
    Rural England
Crenson, M. 149
Creutzfeld Jakob disease (CJD) 95–6,
    100
critical environmental capital 83, 84, 99
cultural influences: human behaviour 41,
    42; satisfiers 45
customers see consumers

dam projects 174
DDT 19–20
debt crisis 174–9, 218
decentralisation 165
decision making: citizen participation
    74, 165; cost–benefit analysis 207–8;
    economic/social responsibility 2;
    incrementalist 151, 152–3; interest
    group influence 144, 145, 147, 148;
    international level 168; mixed
    scanning 151, 153; models of 151–3;
    policy definition 1; rational-
    comprehensive 151, 152; see also
    non-decision making
deep ecology 61
deep environmentalists 59
deforestation 21, 52, 80; Easter Island
    46, 47; public opinion 141; see also
    forests
degradation of waste 21, 23–4, 106
demand 190–3, *190*, *192*, *193*, 194;
    demand management 115; pollution
    taxes 203–4; rational consumers
    215; revealed preferences 206;
    stated preferences 207
demands on government policy-making
    140, 142, 168
democracy 74, 151, 165, 166
Descartes, René 93

developing countries: appropriate
technology 104, 106; Biodiversity
Convention 186; debt 174–9;
environmental capital depletion 210;
forest resources 186–7; Global
Environment Facility 185; Green
revolution 103, 104; health hazards
34; Heavily Indebted Poor Countries
178; manufacturing 110; population
growth 27, 28; privatisation of public
services 182; renewable energy
technology transfers 188; trade 179,
180, 181, 183
direct action 60, 146
discount rates 194–7, 208
dispersal of waste 21, 23, 24
DNA 30–1
domestic insulation 204, 205
Dovers, S. R. 90
Downs, Anthony 140–1
Dror, Y. 153
Du Pont 122–5
Duff, Andrew 61

EA see Environment Agency
Earth Summit (Rio, 1992) 2, 51, 184,
186–7, 213, 218
earthquakes 33, 38
Easter Island 45–7, 49
Easton, D. 141–2
eco-feminism 62
Eco-management and Audit Scheme
(EMAS) 130, 132, 133, 137
eco-system diversity 31
ecocentrism 58–9, 62, 68, 165; deep
ecology 61; ecological economics
213; environmental valuation 205;
rational consumers 215
ecological economics 4, 81, 212,
213–15
ecological footprint 86, 175
ecological modernisation 106–10, 117,
119
economic capital 77–9, 81–2
economic growth 194, 209, 218; Agenda
21 initiative 2, 185; India 28; limits
to growth model 69, 72–3; materialist
values 60–1; sustainable development
74, 75; trade 179
economic instruments 156, 159, 160,
166, 218; 'poluter pays' principle
198–203; population control 162; see
also subsidies; taxes
economics 3, 4, 189–216; capital 7; debt
crisis 174–9; ecological 4, 81, 212,

213–15; environmental 4, 81,
212–13; globalisation 173; resources
8; sustainable development 188; see
also economic growth; market
economics; trade
efficiency: economic 190–1, 193, 194;
energy 99, 157, 159; market forces
193; The Natural Step 135; resource
88, 109, 110, 111, 115
electricity generation: Non-fossil Fuel
Obligation 154–5; privatisation 146,
154, 161; Severn tidal barrage 106,
107–9; see also energy
élite theory 148, 149
EMAS see Eco-management and Audit
Scheme
emissions trading 52, 159, 199–202
employees 120, 127; see also human
resources
empowerment 165, 185
EMS see environmental management
systems
end-of-pipe technologies 110–11, 111,
117, 157
energy: cost-effectiveness 99; domestic
use 2, 159; efficiency 99, 157, 159;
flows 8, 18; limits to growth model
71; The Natural Step 136; prices 159;
renewable 153, 154–5, 188; saving
through recycling 15; service
approach 115–16; solar 8, 10; waste
18, 113, 155; see also electricity
generation; nuclear energy
enforcement: international law 168;
national legislation 156–7
Enlightenment 47, 97
Environment Agency (EA) 88
environmental auditing 128
environmental capital 6–7, 10; Agenda
21 initiative 185; changing behaviour
62; conservation of 217; constant 83,
84, 211; corporate sustainability
targets 133; corporatist policy sectors
148; critical 83, 84, 99; depletion of
39, 64, 137, 210–11; environmental
sinks 15; equitable distribution 165;
extrinsic value 59; multinational
corporations 137; The Natural Step
135; over-exploitation of 44, 45, 101;
quality of life 33–4; Socially
Responsible Investment funds 127;
sustainable development 77–84;
technology 101, 102; tradable 83, 84,
99; tragedy of the commons 4, 50;
valuing the environment 99, 205;

waste concentration effect on 30;
wasteful consumption 49
environmental economics 4, 81, 212–13
environmental groups 119, 144, 145,
146–7; emergence of 60; pluralist
policy sectors 165; post-materialist
values 61; *see also* interest groups
environmental management 86, 217;
CAMPFIRE project 76; local
authorities 132; preventative 110–16,
117, 119; resource renewability 10;
staff development programmes 127;
standards 87, 88
environmental management systems
(EMS) 87, 157, 218; corporate policy
128, 129, *129*, 130–1; Sigma project
133; suppliers 126
environmental problem definition 34–5
environmental responsibility 88, 118,
127
environmental services 6–7, 8, 34–5, 37,
84, 217; environmental space
analyses 84–5; future generations 82;
increased use of 41; The Natural Step
135, 136; quality of life 33; valuing
77, 81; waste concentration effect on
30; wasteful consumption 49
environmental sinks 6, 8, 15, *16–17*, 18,
85, 113
environmental space 84–6
environmentalism 61–2, 64
equity: appropriate technology 105;
economic instruments 159, 166;
environmental economics 213; global
189; inter-generational 75, 77, 84,
102; market forces 194; mutual
coercion 55; The Natural Step 135;
social changes 211; sustainable
development 74, 75, 77, 99, 159, 176;
wealth creation 212; *see also*
fairness; inequalities; redistribution
of wealth
ethical consumerism 215
EU *see* European Union
Europe, sulphur emissions 169–70
European Union (EU): acid emissions
170; agricultural subsidies 181; Eco-
management and Audit Scheme 130,
132; improved industry standards
165; Integrated Pollution Prevention
Control Directive 86–7, 88, 157;
Millennium Round of trade talks 183;
shared sovereignty 168
evaluation of policy 152, 164
exports: debt crisis 176, 178, *179*;

environmental capital depletion 210;
Senegal 171; *see also* trade
externalities 198, 200, 201, 205, 213;
cost–benefit analysis 208; measuring
210, 211; transport 213, 214
extinction 32, 37, 82, 83
extrinsic values 58, 59, 198, 205

Factor Four/Factor Ten 115
fair trade 183, 215
fairness: distributional 73; Malthus 68;
mutual coercion 55; The Natural Step
135; sustainable development 74, 75;
see also equity
falsification 94
FCCC *see* Framework Convention on
Climate Change
feminism 62
fertilisers 103, 104
fertility rates 27–8, *29*
films 63
Finland, forest management 78–9
fish: depletion of fisheries 14; fishing 84,
168, 171–2; populations 11, 12;
resource cycle *9*; restoration of stocks
187
flooding 80
flow (renewable) resources 8–11, *11*, 12,
14; appropriate technologies 105–6;
environmental space analysis 84–5;
Non-fossil Fuel Obligation 153,
154–5
food chain 20, 23
food supply: early agricultural societies
44; limits to growth 69–71, *70*, *71*;
Malthusianism 67, 68
forests 10, 32, 36, 92; depletion 14, 210;
Finland 78–9; Madagascar 80;
Statement of Principles on the
Management and Conservation of the
World's Forests 186–7; *see also*
deforestation
fossil fuels 10, 14, 30, 113; carbon
emissions 21, 22, 28, 53; forestry 78;
The Natural Step *135*, 136; Non-
fossil Fuel Obligation 153, 154–5;
taxes on 109; *see also* coal; global
warming; oil
Framework Convention on Climate
Change (FCCC) 51–3, 185
France: environmental accounting 211;
withdrawal from MAI 182
free markets 180, 190, 193–8, 215; *see
also* capitalism; market economics
free trade 68, 183

Friends of the Earth 144, 145, 146–7
fuel consumption 1–2
fuel protests 142
future generations 63, 64, 75, 165–6;
    discount rates 194–7; environmental
    capital 81, 82, 83; environmental
    valuation 206; natural capital
    entitlement 212; The Natural Step
    135; nuclear waste 196; Structural
    Adjustment Programmes 176; *see
    also* inter-generational equity
futurity: appropriate technology 105,
    106; discounting techniques 194–7;
    market forces 194–7; sustainable
    development 75, 77, 99; *see also*
    inter-generational equity

Gaia hypothesis 93–4
Galbraith, J. K. 33
gas 10, *13*, 154, 161
GATS *see* General Agreement on Trade
    in Services
GATT *see* General Agreement on Tariffs
    and Trade
GDP *see* Gross Domestic Product
GEF *see* Global Environment Facility
gender issues 62
General Agreement on Tariffs and Trade
    (GATT) 180–2
General Agreement on Trade in Services
    (GATS) 182
genes 30–1, 41–2
genetic diversity 31, 104
genetic resources 186
genetically-modified crops 103, 104
Germany, acid emissions 169
Gleckman, Harris 131
Global Environment Facility (GEF) 185,
    188
Global Reporting Initiative (GRI) 133
global warming 21–2, 30, 34, 63, 92,
    217; aviation fuel 136; biodiversity
    32; Framework Convention on
    Climate Change 51–3, 185; human
    influence 38; public opinion 141;
    renewable energy 155; sea-level rise
    25, 27; *see also* climate; greenhouse
    gases
globalisation 172–4, 189; campaigns
    against 59; eco-feminism 62;
    ecological economics 213
GNP *see* Gross National Product
governmental policy making 1, 4, 138,
    139–66; *see also* regulation
Green consumerism 125, 215, 216

Green philosophy 59, 60–2
Green revolution 103–4
greenhouse gases 21–2, 27, 218; dam
    projects 174; Framework Convention
    on Climate Change 51–3, 185;
    resource waste products *16–17*; *see
    also* climate; global warming
Greenpeace 144, 145, 146–7
GRI *see* Global Reporting Initiative
Gross Domestic Product (GDP) 209
Gross National Product (GNP) *179*,
    209–11

habitat destruction 32, 78, 83, 174
Hamburg waste management case study
    112–13
Hardin, G. 4, 49–50, 51, 53–6, 63, 73
health issues: improvements in public
    health 27, 34, 101; pesticides 20
Heavily Indebted Poor Countries
    (HIPCs) 178
hedonic pricing 206–7
Henderson, Caspar 136–7
herbicides 103
hierarchy of needs 43–4, 45, 46, 60
high-yielding varieties (HYVs) 103–4
Hinkley C power station 146–7
HIPCs *see* Heavily Indebted Poor
    Countries
holistic approaches 93, 94
Hortensius, D. 130, 131
human nature 40–2, 68
human needs 35, 43–9, 74, 75, 151, 217
human resources 142; *see also*
    employees
human rights 50, 161, 163, 167
hydrological cycle 11, *12*, 30
hypothecation 200, 203, 204, *204*
HYVs *see* high-yielding varieties

ICT *see* information and communication
    technology
IKEA 136
IMF *see* International Monetary Fund
impact assessment 111, 132
incineration of waste 83, 112, 113
incrementalist decision making 151,
    152–3
India: agriculture 103; carbon dioxide
    emissions 28; population growth 28
industrial production 69–71, *70*, *71*
industrial revolution 47, 60, 97
industrialisation 32, 34, 118
inequalities: access to resources 55, 73;
    appropriate technologies 106; global

172; Structural Adjustment
Programmes 176; sustainable
development 77; trade relations 183;
wealth 173–4; *see also* equity;
redistribution of wealth
information and communication
technology (ICT) 110
information provision 160
Inglehart, Ronald 60–1
inputs *140*, 141–2
insecticides 19, 20
insider groups 145, 147–9, 151
insurance 120
Integrated Pollution Prevention Control
(IPPC) Directive 86–7, 88, 157
intellectual property rights 182, 186
inter-generational equity 75, 77, 84, 102;
see also future generations; futurity
interest groups 143–9, *143*, 152, 165,
166, 168; *see also* environmental
groups
Interface 136
Intergovernmental Panel on Climate
Change (IPCC) 22, 63, 96–7
intermediate technologies 106
international agreements *177*
international bodies *177*
international law 167, 168
International Monetary Fund (IMF) 176
international policy making 4, 167–88
International Standards Organisation
(ISO) 130; ISO 14000 series
standards 130–1, 132, 137; ISO
14001 standard 87, 130–1, 132, 133
internet 116
intra-generational equity 75, 84, 102
intrinsic values 58, 59, 198, 213
investment: discount rates 194–7;
Multilateral Agreement on
Investment 182; Socially Responsible
Investment funds 127
investors 118, 120, 127
IPCC *see* Intergovernmental Panel on
Climate Change
IPPC *see* Integrated Pollution Prevention
Control Directive
irrigation 101, 103
ISO *see* International Standards
Organisation
issue attention cycle 140–1, *141*, 164
issue networks 149, *150*, 165

Jacobs, Jane M. 57
Japan, environmental accounting 211
Jawoyn people 56–7

Johannesburg Summit (2002) 2, 187–8
Jordan, A. 141
Jubilee Movement International for
Economic and Social Justice 178

Korten, D. 137
Krut, Riva 131
Kyoto Protocol 51–3, 185, 218

land 8, 26, 79
landfill waste disposal sites *16*, *17*, 18,
35–6, 83; Hamburg 112; limited
capacity 20; organic matter 24; UK
Landfill Tax 199, 200–1, 203
LCA *see* life-cycle assessment
legislation *see* regulation
LETS *see* Local Exchange Trading
Systems
liability, businesses 118, 120, 122
licensing 157, 198–9
life-cycle assessment (LCA) 15, 111–13,
128, 130, 218
lifeboat model 53–4, 55
limestone *11*, 200, 203
limited liability 118
limits to growth 69–73, *70*, *71*, 88
Lindblom, C. 153
lobbying 60, 120, 143, 144–5, 146, 165
local communities 119, 126
Local Exchange Trading Systems
(LETS) 213, 214–15
logging 178
Lovelock, James 93–4
Lukes, S. 149–51
Lyon, D. 47
Lyotard, J.-F. 98

macro-problems 92
Madagascar, forest management 80
MAI *see* Multilateral Agreement on
Investment
malnutrition 35
Malthus, Thomas Robert 67–9, 73, 104
Malthusian theory 67–9, 163
mangroves 25
Margulis, Lynn 94
market economics 172, 173, 180, 190,
193–8, 215; *see also* capitalism; free
trade
Marxism 148
Maslow, Abraham 43–5, 48, 60, 61, 102,
183, 215
materialist values 60–1
Max-Neef, M. A. 45, 48, 49, 105, 151
Meadows, Donella H. 69, 72, 73

media 140, 144, 145
meso-problems 92
micro-economics 189–98
micro-problems 92
migration, rural to urban 30
mineral resources 7, 12–15, 32, 56–7
mixed scanning 151, 153
MNCs see multinational corporations
modernity 47
Montreal Protocol 125, 218
moral issues, tragedy of the commons
    54, 55
motoring groups 143
Multilateral Agreement on Investment
    (MAI) 182
multinational corporations (MNCs) 126,
    137, 173, 181, 182, 186
mutual coercion 4, 50, 54, 55
Myers, N. 36

national policy making 1, 4, 138, 139–66
national sovereignty 167–8, 181
natural hazards 33–4, 38
natural selection 31, 41, 42
The Natural Step (TNS) 135–7, 135
needs 35, 43–9, 74, 75, 151, 217
Nelson, Jane 120
Nestlé 126
net present value (NPV) 195, 208
networks, policy networks 148–9, 150
NFFO see Non-fossil Fuel Obligation
NGOs see non-governmental
    organisations
Nigeria, cost of environmental
    degradation 210, 211
'no regrets' strategies 98–9, 117, 119,
    217–18
noise pollution 34
non-decision making 149
Non-fossil Fuel Obligation (NFFO) 153,
    154–5
non-governmental organisations
    (NGOs): CAMPFIRE project 76;
    corporate environmental management
    119, 126; development 178; fair-trade
    schemes 183; Johannesburg Summit
    187; UNCED lobbying 184
non-renewable resources 8, 10–11, 11,
    12–15; appropriate technologies
    105–6; environmental space analysis
    85; limits to growth 70, 71;
    reserve/production ratio 13, 13, 14;
    see also fossil fuels
North, R. D. 36
Norway, environmental accounting 211

NPV see net present value
nuclear energy 59, 106, 154, 161;
    discount rates 195; Greenpeace/FoE
    campaigns 146–7; inter-generational
    costs 196–7
nuclear waste 63, 115, 146, 196

oceans 8, 25–7
OECD see Organisation for Economic
    Cooperation and Development
oil 10, 13, 154
open access resources 54, 55, 167, 197
optimism 60, 100
organic waste 23–4
Organisation for Economic Cooperation
    and Development (OECD) 182, 198
O'Riordan, T. 59, 141
outputs 140, 155–6, 161, 168
outsider groups 145, 147, 148, 151
oxygen 35
ozone depletion 34, 92, 123–5, 124, 185,
    218

Paine, Thomas 67
paper 10, 16–17, 78, 114
patents 182
persuasive policy instruments 156,
    160–1, 163, 166, 218
pesticides 19–20, 96, 103–4, 178;
    bio-accumulation 23, 23; forestry 78;
    residues in food 34
petrol 202, 203–4
pluralism 147–8, 149, 165
policy: corporate environmental policy
    making 118–38; definition 1, 2, 38;
    distributional 73; economics 4,
    189–216; evaluation 152, 164; goals
    4, 64, 66, 86–8; governmental policy
    making 1, 4, 138, 139–66; inputs
    140, 141–2; international policy
    making 4, 167–88; outcomes 140,
    161, 168; outputs 140, 155–6, 161,
    168; precautionary principle 99–100,
    117; purpose of 39; science 4, 90–2,
    98, 117; sustainable development 3,
    75, 82, 83–4, 88; 'toolkit' 4; values
    62
policy communities 149, 150
policy environment 140–1, 140, 142,
    168
policy networks 148–9, 150
politics: decision making systems 3,
    151–3; environmentalism 60, 62;
    interest groups 143–9, 143; public

safety 94, 97; representation
142–3, *143*; short-termism 165;
supports 142; sustainable
development 75
'polluter pays' principle 198–205
pollution 7, 35, 41, 92; acidification 30;
agricultural 103–4, 210; BATNEEC
principle 86; bio-degradation of
waste 24; competitive advantage 168;
ecological modernisation 106, 110;
economic approaches to control
157–9, *158*; EU Integrated Pollution
Prevention Control Directive 86–7,
88, 157; Global Environment Facility
185; limits to growth 69–71, *70, 71*;
pesticides 19–20, 178; regulation
157; scientific evidence 98–9; taxes
159, 191, 198, 199–203; tragedy of
the commons 50; transportation 114;
unit prices 191; waste concentration
effect on 30; *see also* acid emissions;
air pollution; greenhouse gases; water
pollution
Ponting, Clive 46
Popper, Karl 94
popular culture 63
population control 161, 162–3
population growth 7, 24–30, *29*, 35, 41;
China 162, 163; environmental space
analysis 85; food supply 44; global
173; limits to growth theory 69–73,
70, 71; Malthusianism 67–9;
projections *28*; sustainable
development 74; tragedy of the
commons 50
Porritt, J. 119
positivism 94, 97
post-materialist values 60–1
postmodern society 47–9, 55, 63, 64,
114, 217; consumption 47, 48–9, 216;
ecocentric reaction against 59;
experts' role in policy-making 117;
free-market principles 215
postmodernism 97–8
poverty 34, 35, 44, 63, 165, 218; Agenda
21 initiative 185; developing
countries 183; economic instruments
159; global wealth inequalities
173–4; Malthus 68, 69; redistribution
of wealth 72, 73; resource scarcity
15; sustainable development 74, 77
power 147, 148, 149–51, *150*
precautionary principle 99–101, 117,
169, 183, 217–18
preservation 58

preventative environmental management
110–16, 117, 119
price-elasticity 193, *193*
prices 14, 190–1, 206–7
primary resources 7–8, 14, 15, 113
privatisation: developing countries 182;
tragedy of the commons 50; UK
electricity industry 146, 154, 161;
waste management 112
producer responsibility 136
product charges 159, 160, 202
production: economic efficiency 191;
environmental space analysis 85;
postmodern society 47, 48, 59, 63;
technologies of 110
propaganda 161
pseudo-satisfiers 49, 105, 151
public opinion 140–1, 164, 183
public safety 94, 95–6, 97
public transport 92, 110, 116, 159,
203

quality of life 7, 8, 33–4, 81, 209
quality management 119, 129

radioactive waste *17*, 196
rainforests 36, 178
rational-comprehensive decision making
151, 152
re-use 15, 105, 114, *114*
reclaimed land 26
recovery of waste 8, 113, *114*, 200
recreation 33
recycling: appropriate technology 105;
ecological modernisation 109, 110;
enforcement of laws 156–7; glass
bottles 202; Hamburg 112; high
resource prices 14; hypothecated tax
revenue 203; The Natural Step *135*,
136; nuclear fuel 115; policy targets
2; secondary resources 8, 15; UK
Landfill Tax 200; voluntary action
160; waste management hierarchy
113–14
redistribution of wealth 72, 73
reductionism 42, 93, 94
reforestation 52, 78
regulation 156–9, 191, 218; air pollution
149; business 119, 120; EU
Integrated Pollution Prevention
Control Directive 86–7, 88, 157;
Framework Convention on Climate
Change 51–3, 185; population control
162; sulphur emissions 169–70;
UK Landfill Tax 199, 200–1, 203;

see also governmental policy making; standards
renewable energy 153, 154–5, 188
renewable resources 8–11, *11*, 12, 14; appropriate technologies 105–6; environmental space analysis 84–5; Non-fossil Fuel Obligation 153, 154–5
reserve/production ratio 13, *13*, 14
resource cycle 7–8, *9*, *10*
resources 6, 7–15; appropriate technologies 105–6; common 54, 55; depletion 11, 12–15, 69, 176, 197; economic capital 81; efficiency 88, 109, 110, 111, 115; environmental space analysis 84–5; governmental policy making 139, *140*, 142; impact assessment 132; international policy making 168; limits to growth 69–72, *70*, *71*; Malthusianism 68; The Natural Step *135*; policy networks *150*; population growth relationship 28; postmodern consumer culture 217; sustainable development 74; taxes 202–3
revealed preference valuation methods 206–7
Rio Summit *see* Earth Summit
risk 99, 100
risk society 97
rivers 6–7, 8, 174
Rose, S. 42
Royal Commission on Environmental Pollution 86
rural areas 30, 106

Sabatier, P. A. 164
St Francis of Assisi 58–9
SAPs *see* Structural Adjustment Programmes
satisfiers 45, 48–9, 102, 215
scarcity: Malthusianism 67; resources 14–15
Schumacher, E. F. 104, 106
science 4, 90–101, 116–17; ecocentrism 58; technocentrism 59; understanding of natural environment 3
scientific knowledge 4, 37, 93, 94, 117, 142
Scottish Environmental Protection Agency (SEPA) 88
sea-level rise 22, 25–7, 96–7
secondary resources 7–8, 15
sectional groups 143, 144
selfish gene 41–2

Senegal, fishing industry 171–2
SEPA *see* Scottish Environmental Protection Agency
service approach 115–16
services: General Agreement on Trade in Services 182; shift to service economy 110
Severn barrage 106, 107–9, 195
shallow ecology 61–2
shareholders 118, 127, 128, 137
Shell 126
Shiva, V. 62
Sigma project 133–5, *134*
silica 12
Simon, H. 152
sinks 6, 8, 15, *16–17*, 18, 85, 113
Sizewell B power station 146–7, 154
small and medium-size enterprises (SMEs) 126, 131, 132
Smith, G. 145
social capital: depletion by MNCs 137; The Natural Step 135; quality of life 34, 81; Socially Responsible Investment funds 127
social exclusion 165
social justice 194
social responsibility 2, 118, 127
social sustainability 133–5
Socially Responsible Investment (SRI) funds 127
socio-biology 41–2
soft technologists 59
soil erosion 46, 80, 210
soil pollution 18, 30, 178
solar energy 8, 10
Southwood, Richard 95
sovereignty 167–8, 181
species diversity 31, *32*
SRI *see* Socially Responsible Investment funds
stakeholders 98, 119–20, 129, 133, *134*, 152
standards 87–8; ISO 14000 series 130–1, 132, 137; ISO 14001 87, 130–1, 132, 133; sustainability 133
stated preference valuation methods 207
Statement of Principles on the Management and Conservation of the World's Forests 186–7
stock (non-renewable) resources 8, 10–11, *11*, 12–15; appropriate technologies 105–6; environmental space analysis 85; limits to growth *70*, *71*; reserve/production ratio 13, *13*, 14; *see also* fossil fuels

strong sustainability 82, *82*, 133, 211
Structural Adjustment Programmes
 (SAPs) 176, 178
subsidies 159, 160, 180, 204–5;
 agricultural 181, 204; public transport
 203
sulphur emissions 169–70, 201
supply 190, *190*, 191, *192*, 215
supports to policy system *140*, 142, 168
sustainability: capital approach 77–84;
 companies 133–7, *134*;
 environmental economics 213;
 environmental space 84–6; measuring
 211; policy development 164; strong
 82, *82*, 133, 211; urban environments
 211, *212*; valuing the environment
 205; weak 82, *82*, 211
sustainable development 66, 73–88, 99,
 164–6, 188, 218; Agenda 21 initiative
 184–5; businesses 119, 120, 137;
 ecological economics 215;
 economic/social responsibility 2;
 equity 159; global inequalities 172;
 Johannesburg Summit 187–8; market
 forces 194; measuring 209–12; policy
 goals 4; 'polluter pays' principle 198;
 progress towards 3; radical policy
 changes 153; SAPs impact against
 176; technology 101–2; trade 168;
 UNCED 184; WTO principles 183
Sweden: environmental accounting 211;
 The Natural Step 135

targets: acid emissions 170; corporate
 environmental policies 130, 131;
 environmental space 85, *85*;
 greenhouse gas emissions 51–2;
 Johannesburg Summit 187; waste
 disposal 2
tariffs 180, 181
taxes 155, 159, 218; ecological
 modernisation 109, 110; pollution
 191, 198, 199–203
TBL *see* triple bottom line
technocentrism 59–60, 62, 67, 75, 165
technology 4, 41, 90, 101–116, 117;
 appropriate 104–6, 117, 218;
 ecocentrism 58; globalisation 173;
 limits to growth model 71; mineral
 prospecting 14, 15; as resource for
 policy formulation 142; sustainable
 development 73–4; technocentrism
 59, 60; unit prices 191
The Natural Step (TNS) 135–7, *135*
TNCs *see* transnational corporations

TNS *see* The Natural Step
tourism 76–7
tradable environmental capital 83, 84,
 99
tradable permits 52, 159, 199–202
trade 168, 178, 179–83, 189, 218;
 globalisation 173; liberalisation 168,
 171, 179, 180; 'polluter pays'
 principle 198; *see also* exports; free
 trade
trade unions 119, 143, 145
tragedy of the commons 4, 49–55, 62,
 155, 160, 167, 197–8
transnational corporations (TNCs) 173
transport 2, 92, 116, 159; ecological
 modernisation 110; externalities 213,
 214; fuel protests 142; globalisation
 173; The Natural Step 136; pollution
 114; taxes 203
triple bottom line (TBL) 120, 126, 127,
 133
Turner & Newall 121–2

UN *see* United Nations
UNCED *see* United Nations Conference
 on Environment and Development
uncertainty 4, 64, 98–9, 117
UNDP *see* United Nations Development
 Programme
UNECE *see* United Nations Economic
 Commission for Europe
UNEP *see* United Nations Environment
 Programme
United Kingdom: acid emissions
 169, 170; BSE crisis 95–6;
 ecological footprint 86;
 environmental space targets 85;
 interest group–government
 interaction 147; Landfill Tax 199,
 200–1, 203; LETS schemes 214;
 local authorities 132; materialist/post-
 materialist values 61; The Natural
 Step 135; Non-fossil Fuel Obligation
 153, 154–5; nuclear energy 146–7,
 154, 161; pension funds 127;
 privatisation of electricity generation
 146, 154, 161; regulation 86, 87, 88;
 Sigma project 133–5, *134*
United Nations (UN) *28*, 51, 184
United Nations Conference on
 Environment and Development
 (UNCED/Earth Summit, Rio 1992) 2,
 51, 184, 186–7, 213, 218
United Nations Development
 Programme (UNDP) 185

United Nations Economic Commission for Europe (UNECE) 169–70
United Nations Environment Programme (UNEP) 133, 171, 184, 185, 211
United States: carbon dioxide emissions 28; improved industry standards 165; interest group–government interaction 147; Kyoto Protocol rejection 53, 185, 218; LETS schemes 214; materialist/post-materialist values 60–1; The Natural Step 135; nuclear energy 196–7; sovereignty 181–2; tariff barriers 181; tradable permits 199–201
Universal Declaration of Human Rights 50, 167
uranium 7, *17*
urban environments 30, 211, *212*
urbanisation 30, 32

Valleron, Alain-Jacques 96
values 37, 55–64, 198, 205, 215
valuing the environment 75, 77, 99; appropriate technology 105, 106; economic methods 205–8; market forces 197–8
vehicle emissions 202
volcanoes 33, 37–8
voluntary action 160, 165

waste 2, 7, 15–24, 35–6; appropriate technologies 105, 106; assimilation 8, 21, 23–4, 85, 106, 113; biodiversity 32; ecological modernisation 106, 109, 110; Hamburg waste management case study 112–13; impact assessment 132; investment in capital infrastructure 197; minimisation 106, 109–11, 112, *114*, 115, 157, 200; The Natural Step 136; population growth relationship 28, 30; postmodern consumer culture 217; regulation 88; resources 8, 15; UK Landfill Tax 199, 200–1, 203; waste management hierarchy 113–15, *114*; *see also* nuclear waste; pollution; recycling
waste-to-energy schemes 155
water: hydrological cycle 11, *12*; metering 116; pollution 18, 30, 34, 178, 185; *see also* rivers
weak sustainability 82, *82*, 211
wealth: creation 178, 179, 180, 183, 212; equity 77; global inequalities 173–4; quality of life 209; redistribution of 72, 73
wetlands 81
whales 37, 59
WHO *see* World Health Organisation
wildlife: forest management in Finland 78–9; reserves 76–7, 83; Severn barrage project 108
wind energy 8, 154, 155
Winter, M. 95
women 62
wood pulp 10, *16–17*, 78
work 47–8, 213
World Bank 30, 185
World Commission on Environment and Development (Brundtland Commission) 73–4, 165, 184
World Health Organisation (WHO) 15
World Summit on Sustainable Development (Johannesburg Summit, 2002) 2, 187–8
World Trade Organisation (WTO) 180–3

Xerox 116

Yorkshire Water 136

zero environmental impact 58
Zimbabwe, CAMPFIRE project 76–7